JAPANESE
TALES AND LEGENDS

DATE DUE

~~2~~			
~~2 '0~~			
MR 1 5 03			
OC 1 9 '07			
DE 2 '08			
MR 1 4 '09			
MR 2 0 13			
AP 1 5 '13			
AP 2 4 13			

Oxford Myths and Legends in paperback

✱

Japanese
Tales and Legends

Retold by
HELEN & WILLIAM McALPINE

Illustrated by
JOAN KIDDELL-MONROE

OXFORD UNIVERSITY PRESS
OXFORD NEW YORK TORONTO MELBOURNE

t, Oxford OX2 6PD

k

ogota Bombay

n Dar es Salaam

tanbul Karachi

d Melbourne

Mexico City Nairobi Paris Singapore
Taipei Tokyo Toronto

and associated companies in
Berlin Ibadan

Oxford is a trade mark of Oxford University Press

© Oxford University Press 1958

First edition 1958
Reprinted 1960, 1964, 1968, 1979
First published in paperback 1989
Reprinted 1990, 1993, 1996

McAlpine, Helen
 Japanese tales and legends.
 1. Tales. Legends. Japanese tales & legends
 Anthologies
 I. Title II. McAlpine, William III. Kiddell-Monroe, Joan
 398.2′0952

ISBN 0-19-274140-3

Printed in Hong Kong

Contents

∽

Contents

EPICS AND LEGENDS

The Birth of Japan

Takamagahara

OF old, there was nothing in the universe but thick, sluggish matter. It was shapeless and formless and stretched to infinity. All was chaotic. Heaven and Earth were mingled like the white and yolk of an egg that had been stirred through countless ages. Aeon followed aeon without change. Then, suddenly, a great upheaval began to take place and strange noises filled the boundless, silent universe, and out of the chaotic mass the light and purer part rose up and spread thinly outwards, while the heavier and grosser elements gradually came together and fell, until there was a distinct cleavage between the two.

The light mass moved steadily upwards. It spread and extended until it completely overhung the solid mass below. Parts of it, as if hesitant and uncertain what to do, still clung together and formed many clouds. But the great expanse around and over them formed a paradise, and it was called Takamagahara—The High Plain of Heaven.

All this while, the heavier mass was still sinking and seemed to have great difficulty in taking shape. Another aeon passed. From the heavenly heights the mass looked huge and black, and it was called Earth.

In this way the Heaven of Takamagahara and the Earth came into being, and with them the legend of the birth of Japan.

II

Izanagi and Izanami

As time passed, three gods were born in the High Plain of Heaven: Ame-no-Minaka-nushi—God of the August Centre of Heaven; Taka-mi-musubi—High August God of Growth; and Kami-mi-musubi—Divine August God of Growth. These three gods looked at the Earth below. They saw no order there; everything was in confusion; and there was no sign of growth or life on the ponderous, inert mass.

They looked and contemplated long and consulted among themselves what could be done to give it order and life.

'Even if we talk until our strength drains from us, we are powerless to change these matters,' they said in despair.

Almost in answer to their anxious wonderings a new race of young, virile gods appeared in the High Plain of Heaven. They were sent by the Lord of Heaven whose divine presence was felt throughout Takamagahara and who, the sages relate, was the very Creator of Takamagahara itself. The young gods joined in the consultations with the three elder deities, and after long deliberations

4

it was decided to send two of the youngest and fairest to the Earth below to subdue the chaos and create beauty upon its turbulent face.

The name of the first young god chosen for this gigantic task was Izanagi, and he was tall and strong as a willow sapling. His companion was named Izanami, and she was delicate in speech and manner and as beautiful as the air that filled the High Plain of Heaven.

All agreed that none were more suited for the task, in bravery and fairness, than Izanagi and Izanami. When the choice was made, the Lord of Heaven called the two young gods to him and said:

'You have seen the chaos of the Earth below us. For too long now it has been in that state—spineless and inert as a giant jelly-fish floating in a sea of space. There is no birth, no growth, no order: just darkness and misery. Go, therefore, my children, to your great labour. Draw together the lighter parts and unite the heavier parts; so dispose them that there is pleasure in the contrast between them. Create order where there is none; and in place of anarchy, let there be laws of growth and development. It is for you, my children, to make for me a worthy and beautiful country of the Earth.'

When the Lord of Heaven finished speaking, he

handed Izanagi a finely carved spear, surmounted by an ornamented, jewelled ball of unsurpassed magnificence and rare beauty. It was no other than the legendary spear, Amanonuboko, one of the greatest treasures of the High Plain of Heaven.

'This spear is my symbol,' said the Lord of Heaven, 'and with it you will achieve all.'

As the two young gods bowed reverently, the Lord of Heaven raised his hand and, in the wondrous space over the High Plain of Heaven, a pinpoint of light appeared. It drifted down like a solitary small circle of foam riding on the sea of the sky. As it came nearer, all saw that it was a white ball of cloud, surrounded by an escort of smaller clouds, whose fringes were as colourful as the High Plain of Heaven itself. It came to rest at the throne of the Lord of Heaven, who said to Izanagi and Izanami:

'This is your carriage and on it you may journey through space at your will. Now it is time for you to be going.'

Izanagi and his beautiful companion stepped on to their carriage of clouds, and all watched intently as it sank towards the Earth below bearing its heavenly passengers.

As they receded gradually from the sight of the watching gods, a rainbow arc of light appeared and curved from heaven to earth in bands of many colours. It was the Bridge of Heaven, and Izanagi and Izanami were bathed in its radiance as they descended.

III

The Bridge of Heaven

Izanagi and Izanami floated downwards until they were level with the highest point of the rainbow. There they halted and, hand in hand, stepped from their cloud carriage on to the coloured bridge. They paused and looked about them. Above was the bright blue of the vaulted heavens; but below all was dark and still. As the rainbow curved from them into the distance, it disappeared in dense fog, and they could see nothing of the floating mass of earth.

They stood gazing about them for some time, then Izanagi said to Izanami:

'We must descend through the fog below, for there lies the Earth and our task.'

Holding each other's hand, and with Izanagi grasping the spear Amanonuboko, they set off down the Bridge of Heaven. Soon they were enveloped in fog so thick that all about them was the darkness of night. But on they walked until they reached the end of the Bridge. Here they paused. Both were greatly troubled, for they could see nothing and feel nothing but the touch of each other's hands.

'Is this then the Earth?' Izanami asked anxiously.

Izanagi did not answer, but plunged his spear into the swirling fog. It sank easily and Izanagi turned and probed with it once more, hoping to find some firm ground for a foothold. But there was none. He plunged it again and again in every direction, but everywhere the fog yielded without resistance to its thrust.

'Alas!' he said dejectedly. 'It is like a jelly-fish as our Lord of Heaven said.'

But no sooner had he spoken, than the fog began slowly

7

to evaporate and light again flowed about them. A tremor shook the spear in his hand and he saw that a clot of mud, clinging to the tip, was slowly detaching itself and falling from it. Then, miraculously, several more clots grew and followed the first: and as the mud grew and fell, it massed together, and water poured from the spear's point and flowed gently round the edges.

As the last traces of the fog dispersed, the sky grew brilliantly light. The two young gods looked down from the Bridge of Heaven. Everything shone with the reflected blue of the heavens, and in the midst of the void beneath them lay an island surrounded by a calm, blue sea.

Holding each other's hand tightly, they watched this divine miracle. Without speaking, Izanagi lifted the spear and probed the island in different parts.

'It is firm, Izanami! It is firm!' he cried excitedly, turning his head and holding the spear before her. 'The divine spear has wrought this!'

They looked again at the island beneath them and both were filled with joy. Suddenly Izanami cried eagerly:

'Let us go and explore its every corner.'

Before Izanagi had time to answer, she had stepped from the Bridge of Heaven on to the warm white, sands of one of the beaches. Izanagi followed after her; and both were jubilant to feel the earth beneath their feet and hear the lapping of the sea among the tongues of rocks which embraced the white beach.

They walked the island from end to end. Everything filled their eyes with joy, and before the expanse of ocean surrounding their new land they stood spellbound.

'It is very small, our island; but it is enchanting, is it not?' said Izanami, but Izanagi only laughed with happiness in reply.

They came to a small plateau, and as they rested side by side watching the heavens above, Izanami suddenly said:

'Izanagi, we are the first gods from the High Plain of Heaven to put foot on this Earth. This is to be our home for always. Let us build a shrine on this plateau, where we can serve the great gods and spend our lives in peace.'

Izanagi was delighted with the idea and said:

'Indeed, indeed! We will build it with our own hands and in its centre we will build a column that will rise to heaven. Then we shall always feel close to our former home.'

They both knelt down and raised their eyes to the High Plain of Heaven, begging the gods to bless and help their endeavours.

Day after day they laboured. The shrine slowly took shape, and the great column in the centre stretched to the skies. When at last their work was done, Izanagi and Izanami made formal preparations for its dedication. They had chosen names for the island, the shrine and the column; and, kneeling, they prayed earnestly that the shrine might be sanctified by The August Lord of

Heaven. To the island they gave the name The Island of Onokoro; the shrine they named Yashirodono or The Palace of Eight Fathoms; and the column, Amano-mihashira or The August Pillar of Heaven.

The naming done, a great peace fell upon them; the air grew still and the tide ceased its sound; the light of evening embraced land and sea. Izanagi and Izanami bent their heads in reverence, for they knew that their prayers had been heard.

IV

The Birth of the Islands

Time passed on the beautiful island. In every direction stretched a vast blue sea of space. Izanagi often stood on the island's highest point, wondering if a visitor from heaven might not honour them with his presence. One day, as he gazed and pondered, clouds of fog and mist began to rise up all around, the seas began to churn and boil, and the waves to dash themselves against the island shores. But even as he watched, the mist began to clear, and a shining roof, for such it seemed, emerged above him. It was the sky, separating itself at long last from the oceans and now filling the vault of heaven with its light. Izanagi was filled with joy at this sight and called loudly to Izanami:

'Come quickly! Come quickly! A new world is coming to birth!' Izanami, hearing his cries, came running to him. Together they looked and marvelled, and the beauties of the island revealed themselves anew to the young couple. Then Izanagi spoke:

'When we were sent from Takamagahara to this lower world, the Lord of Heaven turned it from a spongy mass into this firm and lovely earth. He did so that we might live here and create goodness and beauty in place of

10

chaos. This island, that we have named Onokoro, is fair and enchanting, but it is very small. We must beg Heaven's help to build other larger islands that the world may grow and increase.'

He spoke in a voice full of deep emotion, for he was already filled with the vision of his new creation. Taking Izanami by the hand, he led her to the shrine and there they prayed fervently for blessings on their task. At last they rose from their knees and Izanagi, turning to speak to Izanami, suddenly paused.

'Izanami,' he said, 'to create these other islands, we must become man and wife. We must circle the earthly column, you to the right and I to the left, and when we meet we shall know each other truly.'

So both walked round the column, Izanami turning to the right and Izanagi to the left. When they met on the farther side of the column, Izanami spoke first and said:

'How delightful it is to meet so sweet a youth!'

And although Izanagi replied: 'I am overjoyed at meeting so fair a maiden,' there was displeasure in his voice. Nevertheless they embraced and became man and wife, but there was no joy between them any longer.

In time Izanami bore her divine husband a son, but, to their dismay, it was weak and boneless like a leech.

'This is surely a result of the Lord of Heaven's displeasure with me,' said Izanagi. 'We must not keep this child, it is an ill omen,' and he placed the baby in a boat of reeds and sent it adrift on the ocean.

For many days both were depressed and unhappy. Then one morning the young husband, after consulting the gods in the High Plain of Heaven, said to his wife:

'The gods are displeased because you spoke first when

we met after circling the Heavenly Column. "Man must take precedence over woman," they said, and therefore we must circle it again.'

They both walked to the column, and when they had circled it as before, Izanagi spoke first and said:

'The gods are kind to me in putting in my way so wonderful a maiden.' And Izanami replied:

'I am beloved by the gods who have allowed me to meet so divine a youth.'

They both gazed on each other for a long time; they were filled with a strange wonder and a profound change came over them. They began to feel a sense of unity with the earth around them, and a new love for each other was born in them.

That evening the two young demi-gods—for they had now become part of the earth around them—talked eagerly of the new islands they hoped to create, and both prayed earnestly for help. As they knelt before the shining column Amanomihashira, the sky began to glow with warm, gold radiance. The majestic sun dropped slowly across the arch of Heaven; redder and redder the light fell on the sea, and the waves sent back reflections of purple, rose, and blue. The shadows lengthened and darkened and the sun glowed with deepest crimson as it fell. The island was irradiated with warmth and the column Amanomihashira shone with an unearthly light as it speared the sky. As the sun was quietly swallowed by the depths of the vast ocean, the waves were touched momentarily with its last rays and their crests rose and fell in cascades of stars with the swelling and eddying of the billows. All became dark and a blackness of ebony fell over the sea and wrapped the island in its folds. There was not a sound, and over all was utter stillness. In the shrine Izanagi and Izanami knelt, intent and dedicated.

But the stillness was the harbinger of a coming storm. Soon the ocean began to rise and swell and gradually the waves grew more and more mountainous; the air grew noisy with the sound of the rolling tide and the wind churned the water into angry whirlpools. All through the night the seas roared and thundered; but towards dawn all quieted again and lay still.

When Izanagi and his wife came out of the shrine, they stood transfixed. Before them stretched the long, curving shore of a vast island, and on the far horizon were the shapes of others. In great joy the two set out to view their new domains. From island to island they went, marvelling at each new land; and when they had travelled them all, they found there were eight, and to them they gave these names in the order of their birth: first the island of Awaji; next Honshu; then the island of Shikoku, followed by Kyushu; Oki and Sado which were born as twins; Tsushima; and finally Iki. Together, they were called The Country of the Eight Great Islands, and as time passed they became known as Japan.

More and more islands appeared, and every day Izanagi travelled the land and sea watching over them. Sometimes Izanami went with him, but serving in the shrine took much of her time and she found the long journeys exhausting. One day Izanagi said to her:

'Izanami, my dear wife, now that we have all these beautiful islands around us our work is greatly increased. I see that you are sometimes very tired, and it troubles me. What would you like to do in order to refresh yourself? I beg you not to work so hard.'

Izanami bowed her head and listened submissively while her husband spoke. Then she answered in gentle tones:

'My dear husband, there is nothing I wish to do more than to live here with you in peace and contentment. But now that so many islands have been born I pray that we, too, may bear children for our help and delight.'

Her prayers were answered and in the years that followed many children were born to them. The first was a Sea Spirit, the next a Mountain Spirit, and then in succession the spirits of fields, trees, rivers, and all natural things. Under their care and guidance the islands grew more and more verdant and beautiful. Soon the Seasons were born, and the breaths of the winds and rains brought their changing cycles to the mountains and fields. Everywhere the forests grew thick and dense, and in the groves flocks of birds gathered and sang. Crops and harvests multiplied, and flowers and bushes blossomed in profusion.

Izanagi and his wife lived in the utmost contentment among their family, and when a daughter was born to them, who was a goddess of the Sun, their joy was unbounded. She was a most beautiful and radiant being, and her whole body gleamed and shone with so resplendent a lustre that everywhere she went she filled the darkest air with light and brilliance. Her they named Amaterasu. Soon after a Moon god was born to them, and he too shone with light, next only to the sun in splendour. His radiance was mellow and pale, and he walked among the dark shadows of evening dispersing them with his quiet rays. Him they called Tsukiyomi. When a Sea god was born to them, his silvery green body reflected in changing hues the resplendent light that shone from his brilliant brother and sister.

One day, as Izanagi and Izanami were talking together, a sudden thought struck Izanagi and laughingly he said:

'Izanami, among all our wonderful children, there is not one who truly represents Fire. Our daughter Amaterasu is a Sun goddess, it is true, but we need a Fire god himself to live close to us. Let us pray that he may be born to us.'

The time came when their prayers were answered, and a fiery, powerful son was born to them, whom they named Kagutsuchi, the god of Fire. But his birth proved too much for the mother and her body was partly consumed in the burning heat of her Fire-god child. Alarmed at this unforeseen disaster, Izanagi ran to her side to comfort and tend her. He prepared for her special foods and medicines, but, in her agony and sickness, whatever she attempted to swallow was immediately rejected. And lo! it was from these rejections that the gods and goddesses of metals were born; while from other parts of her body, the god and goddess of Earth issued forth. But now her body completely perished and Izanami passed away. Death had come to the world for the first time, and with it, its earthly companion, grief.

v

Izanagi's Descent to Hades

Izanagi was filled with sorrow and loneliness. Through countless years, he and his divine sister-wife had subdued the chaos and disorder of the earth; they had together created the islands of Japan and brought to them the beauties and order of nature; they had created gods and goddesses to reign over the sea, the winds, the mountains, the sun and the moon, fire, and the earth. Now she was gone and he was alone.

For many days and nights he wept, and his many children around him could not console him, so great was his grief. He called her by name, but there was no answer to his lonely cries. Unable to bear his solitude longer, he resolved to journey to the underworld of darkness, from whence was derived the fiery strength of the child who had caused Izanami's death, to seek her and bring her back to the world of light.

Once he had made his decision, he became calm in spirit; but his sorrow was no less, and tears filled his eyes for his beloved wife. When he started on his journey, he knew the terrors before him. The road was filled with dangers at every point and enemies lay in wait for strangers who ventured towards their land. But his spirit was indomitable, and after a long journey he reached, without accident or mishap, the forbidding, black region of the underworld.

Days and nights he wandered among the drifting shades of that bleak country looking for his wife; but nowhere was she to be found. At last, dejected and exhausted, he was about to lie down to rest, when he saw her graceful figure ahead of him and cried out to her in joy. She turned with a cry of astonishment and rush-

ing up to him embraced him with tears of welcome. Izanagi clasped her tightly and said:

'Come! do not stay here longer. I have come to take
you back with me to our lovely country, for without you
there is no longer joy in our home. There are still works
of divine creation to be completed, for so it was ordained
by our Heavenly gods. How can I do this work without
you? I need you, as wife and helper.'

But Izanami shook her head and her face clouded as
she said:

'I long to go back with you, my beloved husband, but
alas! it is too late. It is no longer possible.'

'Why? Why?' demanded Izanagi, greatly troubled at
Izanami's words.

'I have eaten of the food and drunk of the wine of this
evil place, my august husband, and now I can never
return to earth.'

Izanagi was chilled with dismay, but he answered
boldly:

'I have made this terrible journey to find you. Nothing
will force me to return without you. Nothing! I will do
anything that will make it possible for you to come back
with me to our island home.'

Izanami dropped her head and thought for a long
time. At last she said:

'Forgive me if I seem so lacking in courage. I know the
hardships you have borne for love of me. What, then,
cannot I do for love of you? I will go and beg the Lord
of this country to allow us to return together. All may
yet be well. Only I must beg of you to make me a solemn
promise first.'

'With all my heart. But tell me what it is,' cried
Izanagi.

'Only this. I shall be away some time and you may
become weary. But I must beg you, my beloved husband,

on no account to enter my room. Wait here faithfully, however tempted you may be to look for me or to rest in my house, for my return. Please swear that you will obey me.'

'I give you my solemn word, my dearest wife, and on our dear love I take an oath to wait and not to seek you.'

Satisfied, Izanami left him. Izanagi settled himself steadfastly to wait her return and to observe his sacred promise. The hours lengthened and all around it grew darker and gloomier than before; yet she did not come. Izanagi grew increasingly anxious; he became filled with apprehension and feared some terrible thing had happened to her. Then he became aware of a most noxious smell. Overcome with a strange sense of terror, and forgetting his oath to his goddess wife, he broke off one of the teeth of the comb he was wearing in his hair, and lighting it like a taper, he held it before him and began looking for the source of the smell which had now become unbearable. He traced it to a small doorway, and wrapping his scarf about his face he entered. Holding the taper aloft he found himself in a small chamber. As his eyes became accustomed to the flickering light and shadows, he became transfixed at the sight before him. Lying on a pallet was the sleeping form of Izanami; but despite the gentle breathing which showed she was living, her body was that of a long-dead person. From it came the nauseous stench which had puzzled him, and he gazed in horror at the decaying flesh that only a while before had been so beautiful. Crouched about her head and breast were the eight foul demons of thunder, who regarded the terrified Izanagi with malevolent eyes and with flames belching from their mouths. In sudden panic he turned and fled, dropping the taper in his haste as he dashed out of the dreadful room.

The noise awoke Izanami, and realizing what had

happened, she was filled with anger. Rushing to the doorway she called furiously after the fleeing Izanagi:

'Is this how you keep your promise? Did I not forbid you to enter my room? You have looked at my revolting form and you have destroyed yourself and me! You have put me to shame and now I have no alternative but to hound you to your destruction!'

With that she summoned a crowd of hideous female demons and sent them in pursuit of Izanagi. They were fleet of foot, and though he ran like the wind, they overtook him. He took out the comb from the left side of his hair and threw it to the ground. From it sprang twining grape vines which caught the demons in a close embrace. Then bunches of grapes appeared and the demons fell upon them. For a long time they munched greedily, until not a grape was left, and only then did they remember their quarry, who by now was far away. With fiendish speed the demons began to gain on him again, and he, feeling them at his heels, threw down the comb from the right side of his hair and from it burst a crop of bamboo-sprouts. His pursuers stopped and ravenously began eating them. Izanagi sped on for his life, but the demons, their meal of bamboo-sprouts finished, soon overtook him again. Then Izanagi made water and caused a great river to flow between him and the demons, and so confused did they become that they returned post-haste to Hades. Izanami was enraged when she found that Izanagi had escaped them. She called upon the eight thunder gods who were clustered about her head and breast and bade them hasten after Izanagi and bring him back to Hades.

Together with one thousand five hundred attendant devils, the eight thunder gods set off at top speed, and soon in the far distance espied the running form of Izanagi. In fast pursuit they overtook and surrounded him. Izanagi drew his sword and laid about him right and left. Hateful as the devils were, he could not find it in his heart to kill them; but he succeeded in beating them off so successfully that they retreated.

Worn out and collapsing with exhaustion, Izanagi at length reached a hill called Yomi-no-horazaka in the Province of Izumo, where he found a peach tree. Knowing the hatred in which the demons held the peach, he gathered its fruit as missiles. When the devils returned, he hurled the peach fruit amongst them and they broke up in confusion.

The demons once again returned to Hades and informed Izanami of their unsuccessful pursuit. Her rage and mortification were unbounded, and having rained

upon them threats and punishments, she determined to give chase to Izanagi herself and drag him back into her power.

Meantime Izanagi had found a boulder of immense size. It would have needed the strength of a thousand men to lift it, but Izanagi was endued with such noble qualities of mind and body that he was able to move it alone and thrust it before the gateway to the lower world. He sat down, very weary, but serene and unperturbed, to await what might happen next. It was not long before he suddenly heard the screaming voice of Izanami and saw her form approaching in the distance. He sprang to his feet in astonishment. He had been prepared for anything, but not for Izanami herself coming as a bitter enemy. He remembered their days of joy together, her gentleness and faithfulness, and he was filled with anguish at the change that had come over her. But there was no stopping her approach, and she was only brought to a halt when she found the great stone barring her way.

'What is this!' she cried, infuriated. 'Why have you put this stone against me? Move it at once!'

Izanagi answered quietly that it would remain where it was. Whereupon the angry Izanami battered at the rock with her hands and stamped the earth so violently that its four corners shook and the seas washed high; but the rock remained firm. Then she appealed to him saying:

'Why did you not keep your promise not to enter my room? I was humiliated and you brought great shame to me. Why did you do this? Why? Why? Why?' and her voice rose to an hysterical screech.

But Izanagi answered quietly:

'Here and now I sever our marriage ties. Understand we are no longer man and wife. I have returned to the world of light and I beg you to return without further words to your land of the dead and darkness.'

With these words Izanagi laid down the formula for divorce for all future generations.

But Izanami shook her head and cried angrily:

'If you do this, my beloved brother-husband, I shall retaliate with all my power and destroy every single day a thousand of your people!'

In the same quiet voice Izanagi answered:

'My beloved sister-wife, if you do this, I will cause every single day a thousand and five hundred women to bear a child, and a thousand and five hundred strong people will live.'

Izanagi paused, and then continued:

'Now that the ties which bound us are broken, from this time forth our countries shall be separated. Return in dutiful submission to your land of the dead, and leave me to my task of creating the living world in peace.'

Now there was nothing more for Izanami to say, and bowing her head, she accepted her fate and returned submissively to Hades. From that time all bonds between them and their two worlds were severed.

The Luck of the Sea and the Luck of the Mountain

PRINCE NINIGI was a grandson of the Sun Goddess, Amaterasu, and because of his noble birth, he was known as The Heavenly Grandchild. He was renowned for his benevolence and righteousness, and ruled wisely in the land which had been created by his heavenly forebears. In the course of time he married and two sons were born to him. The eldest was named Hosusori and the younger Hikohohodemi; and as the months passed and season gave way to season, they grew strong in limb and body, and their beauty, their swiftness of foot and their skill in combat were talked of as far as the island outposts of the divine land.

Although alike in form and feature, they each had a remarkable characteristic which made each one unique: the younger brother had the luck of the mountains in his hands, while the elder held the luck of the sea in his. Whenever Hikohohodemi went out with his treasured bow and arrows after the creatures of the wild, he came back with spoils of fur and feathers. When Hosusori went down to the sea armed with his gleaming fish-hook, he returned with a boat filled with fish of the deep.

On a day when they had returned from mountain and sea, the younger brother said:

'My honourable brother, day after day we go out hunting, and day after day we return with the same spoils of sea and mountain. Such monotony is a bad thing. Let us just for one day exchange our luck. I will lend you my fine bow and arrows and you in turn lend me your gleaming fish-hook.'

After a long silence the elder answered:

'No! I cannot agree to such a thing.'

'But, elder brother, what harm can it possibly do? And think what an interesting and unusual experiment it would be!' urged the younger.

But the elder brother remained obdurate in his refusal. Once more the younger brother addressed him:

'Elder brother! Go to the mountains but once. Take with you my bow and fit to it one of my swiftest arrows; draw the bowstring back from its supple arc and you will hear a sound more beautiful than any in the world as you release the feathered arrow to its target. Birds of every feather and beasts of every fur will fall to you and be your spoil. Come, try it but once, I beg you.'

Hosusori relented and agreed to the exchange, but only for as long as one day's journey of the sun. He handed his fish-hook to his brother with many exhortations as to its care and use, and Hikohohodemi in turn surrendered his bow and arrows.

The younger brother went down to the sea. All day he fished, but not a scale fell to his line. Again and again he sent the gleaming hook into the depths and again and again he drew it up empty. As day closed he flung the line far out for the last time. There was a mighty bite and he drew it in with all his strength. Then something snapped, and to his dismay he brought the line up limp and dangling, and with the precious hook gone.

Meanwhile Hosusori went to the mountain in quest of bird and animal. All day he tramped but nothing could he see. Evening closed over him. His limbs were torn and bleeding and in dejection he murmured:

'What a fool I have been to exchange my sea gift for this worthless gift of the mountain. Never will my line and hook leave my hands again,' and so saying he descended the mountain and went to the sea, where he found his younger brother sitting on a rock by the shore gazing into the darkening depths.

'What was your luck with my line and hook?' asked the elder brother as he approached.

'The luck of the sea is not my luck,' murmured the younger in reply.

'And the luck of the mountain is not my luck,' said the elder vehemently. 'All day I have tramped and tracked, but not a living thing of the hills did I see. My only spoils are torn and bleeding limbs. But this is a lesson not to interfere with luck. We have hurt our heavenly father who bestowed on each of us this gift as our birthright. Now here is your bow and arrows. Return my line and precious fish-hook to me and let us never again be so foolish.'

When the younger looked up and told his brother his hook was lost, the elder's anger was the anger of a storm. He refused to listen to his brother's protestations of regret or his pleas for forgiveness, and only insisted that no matter how, the precious fish-hook must be returned. And flinging the bow and arrows to the ground the irate Hosusori disappeared into the darkness which had fallen about them.

Hikohohodemi was deeply grieved by his brother's outburst. All night long he remained on the rock by the shore perplexed and unhappy, and when dawn came he decided to sacrifice his sword in fashioning new fish-hooks for his brother. In the days that followed he made five hundred hooks from the metal of his sword, and these he took and offered to his brother as recompense for the one lost in the depths of the sea. But the elder brother thrust them aside and demanded once again that his own should be found and returned.

The younger brother, more deeply wounded than ever by his brother's harsh nature, took to haunting the seashore night and day in his despair, searching vainly in the hope that the hook might be washed up by the tide. But his searchings came to nothing, and he cast himself down on the sands and fell to bitter weeping. For long his tears had drenched his sleeves, when he felt a touch on his shoulder, and looking up, he saw the benign, wrinkled face of an old man.

'Surely you are the younger son of the Heavenly Grandchild, are you not? Why do you weep so bitterly?' And so saying he sat down by the young man.

Hikohohodemi was greatly surprised, for he had believed himself alone. But he dried his tears and answered:

'Yes, old sir, you are quite right. I am Prince Hiko-hohodemi, the younger son of Prince Ninigi. Please tell me who you are?'

The old man smiled and said:

'My name is Shihotsuchi and I am Lord of the Tide. What makes you grieve and weep so much? Can I not help you?'

'If, sir, you are indeed the Lord of the Tide, then you must have a knowledge of the sea and its treasures as deep as the sea itself?' said Hikohohodemi.

'That is true,' replied the Lord of the Tide.

Thereupon Hikohohodemi poured out the cause of

his sorrow and the old man listened intently and sympathetically. When the Young Prince had finished, he said:

'You may look for a year and a day, and never will you find the hook. But I can help you.'

With these words he drew from his bag a black comb which he flung to the ground, and from the sand where it fell sprang a grove of bamboos. The old man cut the bamboos and from them fashioned a basket without a seam or a chink in it, and when it was finished he said:

'In this basket you can sail to the bottom of the sea where lives the God of the Sea. It needs no steering for it knows the road well of itself. It will take you direct to the great gates, where grows a cassia tree whose many branches shadow a well by its side. Climb this tree and seat yourself in its top, and by and by the daughter of the Sea God will come and give you counsel. Go now! And may the Heavenly Gods protect you in your enterprise.'

Any doubts the young prince had about these curious instructions and the even odder boat were dispelled by the gentleness and benevolence of the old man. Without hesitation he stepped into the basket, and the Lord of the Tide set it on its course across the sea with a last farewell.

The basket moved of its own accord and soon Hikohohodemi fell asleep. When he awoke the basket had come to rest on a shining yellow strand, and from the sands rose a high gate of exceeding beauty with towers of vermilion coral and a curved, sloping roof of lapis lazuli tiles. Behind lay a palace, and in front of one of the towers of the gate grew a cassia tree with a well beneath its fanning branches. It was all as foretold by the old Lord of the Tide, and for a moment the young prince was spellbound. Then, remembering the counsel of the old man, he climbed the cassia tree, and in its top he sat and waited.

27

After some time, the gates opened and a serving maid appeared carrying a jewelled vessel. She approached the well and was about to draw water, when she saw a light on its surface. Looking up, she beheld to her astonishment the beautiful face of Hikohohodemi peering down at her through the leaves.

'Be not afraid,' said the young prince, 'I have come on a long journey. I am thirsty. Kindly give me some water.'

Enchanted by his looks and voice, the serving maid filled the jewelled urn and handed it up to the handsome stranger. After sipping a little, he dislodged a jewel he was wearing round his neck, and putting it into his mouth, spat it into the urn; then he gave the urn back to the serving maid. When she saw the jewel lying in the bottom, she reached for it; but great was her astonishment when she found that it stuck fast to the urn, and she could not move it. Hurriedly she returned to her mistress, Princess Toyotama, the daughter of the God of the Sea, and handed her the urn. Gazing at the jewel, the princess said:

'Is there some stranger at our gates?'

'There is, most Honourable Princess. He is as handsome and noble as the dawn and sits atop the cassia tree. He asked for water and I gave him some, but he sipped a little only and then spat this jewel into the urn, where it remained fastened,' replied the serving girl.

'I will see this stranger for myself,' said the princess, marvelling at what she had heard.

She went out of the gate to the well. She looked into the water, but when she saw the beautiful image of the stranger gazing at her, she was overcome with shyness and hastened back to the palace with her veils held before her burning cheeks to tell her father. She said:

'There is a stranger before our gate. His form is most

gracious and his face most noble. He is among the branches of the cassia tree. But I do not know his name.'

The God of the Sea called his attendants and bade them open the gate for him. He went out and looked at the stranger in the tree and said:

'Yes! I know you. Are you not the son of the Heavenly Grandchild? But why are you sitting like a bird in the tree? Pray come down and enter our home.'

The young prince came down from the tree and entered the palace, where the Sea God prepared for him a dais of eight-fold silks and eight-fold sealskins. A feast was prepared of delicacies from the eight corners of the sea, and the princess herself served him with fragrant wines.

Tide after tide flowed and ebbed and Hikohohodemi, amongst these new delights, thought less and less of his mission, until at last it faded completely from his mind. The God of the Sea gave his daughter to the prince in marriage, and they lived in happiness for three years.

At the end of the third year, Hikohohodemi remembered the past and grew greatly troubled and preoccupied. His wife became anxious at the change in her husband, and hearing him sigh heavily one day, she ran to her father and begged him to find out what was ailing her beloved. The God of the Sea went to Hikohohodemi and said:

'We are deeply concerned with the unhappiness which has suddenly come upon you. For three years our lives in my palace have gone by unspoiled by sighs, but now you sigh each day and we know not the reason. I beg you to confide in me. Whatever your trouble is, we shall find a way to help you.'

Hikohohodemi was ashamed to have excluded his wife and her father from his confidence, but now he told them the story of the lost fish-hook and his brother's demand to have it returned.

'Of late,' he continued, 'my thoughts have been turning more and more to my native land, and I have remembered my mission and my promise to my elder brother, and I repent that I have been so unfaithful to my trust.'

The God of the Sea bade him cease worrying and assured him that the hook would soon be found. He thereupon called to his gate the fish of the sea, both large and small, and asked of them:

'Have any of you seen or taken a fish-hook?'

But none of them had seen it or taken it.

'But,' they all replied, 'the bream has had a sore throat for some time past and has not come. Perhaps it is the hook which is causing it.'

The bream was summoned to appear, and when her mouth was examined, the fish-hook was found inside it. It was removed and washed and presented to Hikohohodemi, whose only wish now was to return the hook to his brother as soon as possible, although the thought of leaving his beloved wife was more than he could bear. He said to the God of the Sea and his daughter:

'I thank you with all my heart for finding the fish-hook which had brought me so much unhappiness. Now my promise can be fulfilled. I beg you to give me leave to return to my native land. I shall come back to you with the swiftness of my heart's affection.'

The princess wept and her tears drenched her sleeve. Deep was her devotion to him and her heart ached at the thought of parting. But she understood her husband's desire to complete his mission and besought her father to allow him to return to his own country. The God of the Sea said:

'Younger son of the Heavenly Grandchild, go and return the fish-hook to your elder brother. Take with you these two rare jewels. One is the Jewel of the Flowing Tide; the other the Jewel of the Ebbing Tide. When you produce the Jewel of the Flowing Tide, the oceans will rise and flow and the hills will be submerged; when you produce the Jewel of the Ebbing Tide, the waters will recede and vanish from ocean, lake and river. Now, when you return the fish-hook to your elder brother, do so with your back towards him and, at the same time say, "A hook of poverty, a hook of ruin, a hook of wretchedness", and then fling it to him. If he is angry and wishes to do you harm, produce the Jewel of the Flowing Tide and submerge him. If he repents of his ill nature, produce the Jewel of the Ebbing Tide and save him. If you vex him in this way, then will he become submissive and cease to oppose you.'

He handed the young prince the jewels and then called together all the crocodiles of the sea and said:

'The son of the Heavenly Grandchild is about to proceed to The Upper Land. Who among you can carry him the swiftest?'

Each of them answered according to his size and strength, and all begged for the privilege. But one that was a full fathom in length said:

'In all the oceans there is none so strong of body or swift in speed as I. I can carry the son of the Heavenly Grandchild in less than a day's journey of the Sun.'

Hikohohodemi took a sorrowful farewell of his wife, and then the God of the Sea placed the young prince on the back of the great crocodile, urging him to remember his words.

The crocodile sped through the sea with the swiftness of lightning and reached the Upper Land in less than a day's journey of the Sun. When the crocodile was about to return, the young prince untied the small dagger that was attached to his belt and, fastening it round the crocodile's throat, said:

'You have done well. Take this dagger as a token of my thankfulness for your service. And now farewell.'

The young prince set off at once for his brother's house and returned to him his fish-hook in the manner and with the words as prescribed by the God of the Sea. Immediately the elder brother, delighted at having his precious fish-hook with him again, went down to the sea to fish, but he drew nothing but a bare hook. Day after day he cast his line, but always it was an empty hook he drew from the depths.

In fury against his younger brother, he assembled an army and led it threateningly to his brother's house. When he was about to attack, the young prince brought forth the Jewel of the Flowing Tide, and waters rushed from all directions. The elder brother and his men ran swiftly to the nearest hill, but the waters rose and were about to submerge them, when the elder brother cried:

'Pray save me, save me! Forgive me, younger brother, and call away these mounting seas from about me. I beseech you, spare my life.'

The young prince brought forth the Jewel of the Ebbing Tide, and the waters receded and the sea and the rivers were restored to their normal flow. When the elder brother saw that his younger brother was possessed of these marvellous powers of the sea, he came to him and with bowed head said:

'Henceforth, my brother, I shall serve thee faithfully.'

In the palace under the sea, the daughter of the God of the Sea was with child and pined in solitude for the return of her husband. When he failed to come, she took the journey to the Upper Land accompanied by her younger sister, and a great splendour was cast everywhere about the waters. She presented herself before her husband, who was overjoyed to see her, and she said:

'I am with child. The hour of my delivery is near. I have come to you here, for it is not becoming for a child of a heavenly deity to be born on the sea plain.'

The young prince immediately built a hut by the edge of the shore for her delivery, and thatched it with cormorants' feathers. But before it was complete, her urgency became great, and she went inside saying:

'When your handmaiden is in travail, I pray you do not look upon her.'

But the young prince thought her words strange and his curiosity was aroused. He sat by the waves until the tide flowed high; but he could not restrain himself, and he went to the hut and looked upon her at the moment of her delivery. To his horror, the daughter of the God of the Sea had turned into a crocodile eight fathoms long and was writhing in her labour pains. The young prince turned and fled.

When the child was delivered, the Princess of the Sea was ashamed and angry, for she knew her husband had looked upon her. She came to him and said:

'You have seen me and now my secret is no more. I have travelled to you by the paths of the sea, and by the paths of the sea together we would have returned. Always I would have had the land and sea communicate with each other by these friendly lanes. But I am now disgraced and I must leave you and for ever sever the sea and the land.'

With these words, she wrapped the child in rushes and entrusted it to the care of her younger sister. Then she turned to her husband and made a song for him, for her heart was still full of him, and closing the boundary of the sea, she passed from his sight.

Hikohohodemi made a song in reply as the seas enfolded her. He lived many, many years and sang his song to the sea each night: but there was no answer save the sound of the waters lapping on the strand.

Tales of the Heike
From a twelfth century Japanese epic

I

The Secret Meeting at Shishi ga Tani

KIYOMORI had retired from military service and was living in seclusion as a monk. He had been leader of the Heike clan and had held a high place at court. Torn between the duties these positions laid upon him and his longing to escape all earthly pleasures and spend the remainder of his days in prayer and contemplation, he finally made his decision during a period of peace between the warring factions of the clans. He laid aside his sword, shaved his head, and exchanged the atmosphere of court intrigues and the turmoil of the battlefield for the peace of the temple. Despite this, everyone knew, and none better than the cloistered Emperor Goshirakawa, that Kiyomori, even in retirement, was a force to be reckoned with in any questions relating to the state or the army.

Goshirakawa had many favourites at Court, including Narichika—a distant kinsman of Kiyomori, though of another clan—who longed for a position in the army.

Taking advantage of Kiyomori's retirement, Narichika immediately set about preparing the way for his appointment as a commander. His subtle intrigues soon bore fruit. He gained an audience with the Emperor and pleaded his cause with eloquence and force. Goshirakawa was not unfavourably disposed towards the idea, but thought it unwise to act too hastily in promoting Narichika, for fear of rousing the powerful Heike.

The position sought by Narichika was one of supreme importance, and Kiyomori had long decided that it must at all costs be held by one of his own sons. When he heard of Narichika's intrigues and his audience with the Emperor he was beside himself with anger. Narichika's appointment would present grave danger to the Heike, and Kiyomori acted immediately. Leaving his mountain retreat, he went at once to the cloistered Emperor, vigorously opposed any suggestion of Narichika assuming a command in the army, and openly demanded the position for his son Munemori on the grounds that the country would again be plunged into civil strife if Narichika were placed in such an important and strategic position of command. But Goshirakawa, secretly favouring Narichika, failed to be moved by Kiyomori's grave warning and Kiyomori left, angered and bitter, without having gained his point.

The news of Kiyomori's visit to the palace reached Narichika. Realizing that the refusal to appoint Kiyomori's son showed where the Emperor's preference lay, he decided to press for his own appointment without delay, and accordingly asked for an audience.

'Kiyomori is ruthless, arrogant and selfish,' argued Narichika to the Emperor. 'He is not interested in the peace and the welfare of the country, but only in retaining power over the army within his own family, and is even prepared to see it go to his doltish, obstinate and

weak-willed son. For these reasons he is hated by factions of his own warriors who only await the opportunity to rise in revolt against him. Your Divine Majesty
has long wished to see the Heike humbled. Now the
opportunity has come. I can assure your August Majesty
that I can win the dissatisfied section of the Heike army
to me and destroy Kiyomori and his followers for good.'

Goshirakawa was greatly impressed by Narichika's
argument, and as it was his secret wish to see Kiyomori
overthrown and humbled, he granted Narichika's request and appointed him a commander.

Narichika lost no time in seeking out several of his
closest allies, including Shunkan, an elderly priest,
Yasuyori, an official of the Government, and his own son
Naritsune, who had already distinguished himself as
a warrior in a number of encounters. With these three
and a number of other dissenters, who hated the Heike
and were prepared to support him, Narichika arranged
to hold a secret conclave. The priest Shunkan, who
had a wide reputation for wisdom and learning—qualities
which hid a wily and cautious nature—had a small house
in the country at Shishi ga Tani, and it was agreed that
all should assemble there in the utmost secrecy.

On the night agreed for the conclave, each of the group
made his way separately to Shishi ga Tani and thence to
Shunkan's cottage. At a pre-arranged signal, a screen
slid back and the figures of the conspirators disappeared
into the dimly lit interior. Once they were gathered
inside, Shunkan led them into an inner windowless
room, where their dark outer robes and soft voices were
cast off immediately. Shunkan had prepared a lavish
dinner. As the wine flowed freely, pent up feelings
against Kiyomori and the Heike were given full expression, and all vowed to give their lives, if necessary, in
bringing about his overthrow.

As the evening proceeded, spirits rose high. Wine cups were exchanged and the health of each one in turn was drunk. Narichika passed his cup to Shunkan. He was about to fill it from the small porcelain jar, when the neck suddenly cracked and the head fell cleanly off. For a moment there was silence. Then Shunkan called out loudly:

'An omen! An omen! So may the heads of the Heike fall!'

At this more cups were exchanged, and in their intoxication they felt that victory was assured. The drinking continued throughout the night, and the morning found them all asleep, their bodies lying where they had fallen in drunken stupor.

But there was one who remained awake, a friend of Shunkan who detested Narichika and who was in fact a follower of Kiyomori. He had been careful to keep his drinking under control so as not to lose his wits, and looking round the sleeping, drunken bodies, he decided that it would be safe to leave. Silently and swiftly he slipped from the cottage and went at once to Kiyomori to whom he betrayed the plot in all its circumstances.

Kiyomori acted swiftly and with decision. He sent couriers to all quarters with urgent instructions to each captain to arm his men and seize all those who sympathized with Narichika. When Narichika and his fellow-plotters returned to the city, they found it in a state of siege and upheaval. Narichika, already suspicious at the disappearance of Shunkan's friend, realized at once that he had been betrayed and decided that their only hope of escape was to take refuge in a hide-out in the city.

In the meantime, Kiyomori, furious with anger at their treachery, rushed straight to Goshirakawa and asked to be released from his monk's vows. He was determined to don his armour again and lead his men until the opposition had been wiped out. Reluctantly the Emperor consented, but he refused to listen to the report of Narichika's activities and it suddenly dawned on Kiyomori for the first time that the Emperor might, in fact, be in sympathy with the traitors. His suspicion aroused, he was more determined than ever to annihilate Narichika and his sympathizers. Round the coast he whipped up the fighting boats and in the city he sent out his warriors to search every house. Narichika, his son, Shunkan and Yasuyori were soon tracked down to a temple in the outskirts where they were taken completely by surprise. They were forced to surrender and were brought before Kiyomori.

With the Emperor, Kiyomori sat in state, dressed in the full armour and noble helmet of a general. As the prisoners were brought in, his face became suffused with an angry red, while his eyes were almost closed with hatred. The Emperor was loath to condemn his favourite, Narichika, and listened sympathetically to the prisoners' pleas for forgiveness. Kiyomori sat in silence; he saw that the Emperor was determined to make their sentence as light as possible, if not to save them, and whatever he said at this juncture would be a waste of words. Yet his scornful, wrathful face boded ill for the prisoners.

When the prisoners were returned to their cells to await further investigation of the plot, and the Emperor had withdrawn, Kiyomori at once called his captains together and said:

'I have always deferred to our August Emperor, and in honour and respect for him I yield to none. But it is my duty to inform you that he is in league against us. He has lent a willing ear to Narichika's intrigues; he has made him a commander of the army; and even now, when Narichika's treachery has almost plunged our country

into civil strife again, our Emperor is manœuvring to have the traitors spared. After considering carefully all his recent actions in relation to Narichika, I am bound to conclude that our august Lord is part of the plot to overthrow the Heike, and we must not suffer it. My remedy is drastic and unavoidable; if we are to save ourselves and prevent bloodshed in our country, we must force our way into the palace and take the Emperor prisoner.'

Silence hung over the room at these words. Every man present was fully aware of the gravity of the situation and that immediate action was needed to keep it under control. But to force an entry into the palace and put the Emperor under arrest was another matter. How would the people react if it became known, as it quickly would, that the Son of Heaven's palace had been violated and that he had been taken prisoner? Would they not rise in revolt against the Heike? Yet the alternative was certain bloodshed, with the possible overthrow of the Heike and death to themselves. Their minds were quickly made up: they would back Kiyomori, whatever the outcome, in his audacious plan to take into custody the sacred person of the Emperor.

Having won the approval of his captains, Kiyomori ordered Munemori to assemble a thousand picked warriors and to advance on the palace of the Emperor at once.

'Treat the Emperor with every courtesy,' commanded Kiyomori. 'Do not violate his person in any way. Explain to him the reason and necessity of our action and say that he need have no fear for his life or property. As for the others—stamp them out! Behead everyone instantly! Above all, see that Narichika and his spawn Naritsune do not escape your worthy swords.'

While Kiyomori was speaking his eldest son, Shigemori, dressed in ordinary kimono, entered the room and listened to his father with clouded brows. When he finished speaking, Shigemori, unable to contain himself, cried out:

'Father, wait I beg you! I heard your commands to Munemori and implore you to think again before you carry out this rash and ill-conceived act.'

Munemori thrust himself between his father and his brother, and eyeing Shigemori contemptuously, cried:

'What, you a soldier! And questioning your commander's word? Where is your armour? Have you suddenly acquired a woman's heart?'

But Shigemori looked calmly at his father, and Kiyomori, who loved and trusted his elder son and had often reason to be thankful for his wise and thoughtful counsel, remained silent before the unflinching gaze of his son.

'Munemori, you speak only from anger and do not weigh the reasons for your acts,' said Shigemori, turning to his younger brother. 'Yet I can understand, for you have always been impetuous and unthinking in all your undertakings.'

Then with deep earnestness he addressed his father and the assembled company.

'But you, my father, where is your patience and wisdom? Have you reflected upon the consequences of this rash act? You know what such acts have led to in the past. The memories of them are too close for you to have

forgotten so soon. They have led to bloodshed and
violence and a devastated country. Today we have
peace—the first we have enjoyed for many years. Old
scores and hatreds are being slowly erased from men's
minds. Why do you wish to stir them up again and pro-
voke an even worse state of war than before? I warn you
all that if you do this outrageous act, the name of the
Heike will suffer a blot which centuries of spilt blood and
tears will not wash out. Can you, my father, who have
dedicated yourself to a holy life, think that by laying
aside your priestly robes you can lay aside your vows?
Did you not put away your sword and shave your head
because you had turned away from bloodshed? Why
then take to it again and ravish the peace you, and we all,
have longed for?'

At his elder brother's words, Munemori dropped his
head and feelings of shame rose in him. Kiyomori
turned away, and taking up his priest's robes, stood
apart fighting against the emotion which threatened to
overwhelm him. In his heart he knew that Shigemori
was right, but his anger against the Emperor and the
traitors was such that it could not be easily pacified.
Pulling his priest's robes over his armour, he turned
back into the room and sat down, but remained silent.
He looked again the priest he had been, but where the
robe fell apart the armour underneath was revealed.
Looking at this contrast, Shigemori again addressed
him:

'My dear and honoured father, forgive the violence of
my words. Now I speak to you as soldier and priest—
spare Narichika and his son, for they have only acted in
accordance with what they believed their duty.'

But Kiyomori's face grew stern again and he replied:

'My son, what I do, I do for the honour of the Heike
and for the good of you and your brother. This Narichika

is a traitor and Naritsune will follow his father in everything. If they are allowed to live they will only create
trouble for us and for our family in the future.'

'Father, though they are traitors, they are also con-
nected to us by family ties. We cannot take their blood
on our heads. And remember, Narichika is a close
favourite of the Emperor. Exile him and his son, but
spare their lives.'

'Shigemori, I know well the nobleness of spirit that
prompts you. But the Emperor has shown that he is in
support of these traitors and we can never be safe while
he and they are free to plot against us. Better by far to
exterminate Narichika and his followers and to place
Goshirakawa under guard where he cannot harm you
or the Heike.'

Kiyomori spoke quietly. Every word had been
earnestly considered and it was evident that he was
deeply moved. His captains sat in silence; their heads
were bowed and no one moved. Then Shigemori spoke.

'In this world we are dedicated to four loyalties: the
first is to God, the second is to Country, the third is to
Family, and the fourth is to Man. Those who do not
fulfil their obligations to these four are not worthy to live
amongst us. The Emperor is a child of the divine Sun
Goddess. He is as God to us on earth and he rules us and
our country. In giving our loyalty to God we are also
giving our loyalty to the Emperor, and in giving it to
the Emperor we are giving it to God. God, the divine
Emperor and our sacred Country are one and the same
loyalty. Our duty to them is our first and sacred obligation. What is done against the Emperor is against God
and Country. To violate this sacred obligation of loyalty
will bring shame and defeat to the Heike. I, being a
soldier, must protect my Emperor. And if you persist
in your plan, I must fight against you.'

'Against me, your father? You would fight against me, Shigemori?'

Shigemori knelt before his father and said in anguish:

'Father, don't you see? As a soldier I must fight for my Emperor. But as your son I cannot turn against you. What am I to do? If nothing I say can move you, and you are determined to attack the palace, then, before you send your soldiers there, I beg you to grant me this last request—behead me with your own sword!'

As he finished speaking, Shigemori wept bitterly, and unashamedly the tears fell from the eyes of Munemori and the gathered warriors as they wept in sympathy. Kiyomori laid his hand gently on his son's shoulder and in a sad, resigned voice said:

'Shigemori, you have defeated me. I can do no more. I wished to protect you and our family. But you are right. Do not grieve longer. Narichika and his son will be spared and the Emperor will be unmolested. Do with the prisoners as you will.'

Kiyomori, without glancing once at the assembled company, wrapped his priest's robes more firmly around him and went from the room.

Left to themselves, Munemori and the captains gathered round Shigemori. But he did not spare them. He blamed them, and in particular Munemori, for not dissuading Kiyomori from such wild actions. He reminded them that Kiyomori was getting old and that he needed guidance and advice from those who would succeed him. If the power of the Heike was at stake, so also was the peace of the country. But both could be preserved if wisdom and good counsel prevailed.

Without saying another word, Shigemori left the room and went straight to Kiyomori. With his permission he proceeded to the Emperor and urged the exile of the traitors. Goshirakawa, after listening to Shigemori's reasoning, was wise enough to see, despite his desire to protect Narichika, that punishment was necessary to appease Kiyomori and the Heike, and agreed to the banishment of Narichika to Bizen—an isolated part of the mainland. For Narichika, this was a sore trial, but somewhat mitigated by the fact that it was still a part of the mainland, and should matters ever be reversed, he was comparatively near at hand. But for his son Naritsune he was deeply anxious; for Naritsune, the official Yasuyori, and Shunkan the priest were condemned to banishment on the distant and unfriendly island of Kyushu until death should release them from their exile.

II

The Poem from the Sea

On the ship that conveyed them to their place of exile far beyond the southern shores of Kyushu, Naritsune and his two companions, Yasuyori the official and Shunkan the priest, spent many weeks of manacled confinement. They were held in the deepest hold of the vessel. The heat was oppressive and the air foul. With only a meagre ration of rice and water each day, they suffered from burning thirst and hunger throughout the long voyage. Their ankles and wrists were raw with the rubbing of the chains, and each roll of the ship brought with it new agonies of pain. Their thoughts dwelt on their immediate sufferings and were dead to the future and the fate in store for them. And happy for them it was so; for such a knowledge, added to their present miseries, would have rendered insupportable a punishment already rigorous beyond their worst imaginings.

When at last the ship rode quietly into a bay and the rhythmical noises of the oars had ceased, they were unchained and dragged to the deck. Their eyes, so long

accustomed to the darkness of their prison, strained and watered in the bright sunlight, and only dimly did they perceive the desolate shores of their island of banishment lying just beyond. Forced overboard and harried by the merciless guards, they floundered and stumbled through the shallows to the shore, where they fell, too exhausted to be aware of the departure of the guards, who cast scarcely a backward glance at the three figures huddled at the water's edge.

They lay for a long time without speaking or moving, each only conscious of the relief of being freed from the ship's ghastly hold and its tormenting chains. Slowly they recovered, and when they began to look around them, panic and dismay overwhelmed them. Sulphurous yellow rocks and boulders strewed the landscape everywhere. Neither tree, root, nor grass relieved the desolation of the scene. The sun beat down mercilessly upon the parched land, and the dazzling expanse of sea stretched unchanging in every direction. Fumes from hot sulphur springs which were scattered about the island spiralled upwards to the sky, and the silence was broken only by the lap of waves on the rocks. Shocked and aghast at these surroundings, they wandered in the hope of finding habitation or vegetation: but when night fell, only the same unearthly scene presented itself. Hungry, thirsty, and exhausted, they lay together in the shelter of a rock to sleep as they could till daylight came.

Weeks, and then months, passed. Somehow they created a bare existence for themselves out of the island wasteland. With improvised hooks they caught fish from the shallows, and no rock-clinging herb of the sea or land was too unpalatable for their everlasting hunger. Less and less did they resemble human beings from a civilized world, and more and more they became like some aborigines of their fierce wild environment. Their eyes ailed from staring at the pitiless glare of the sulphur-corroded rocks; their hair grew long and matted to their shoulders; and the flesh fell from their thin undernourished bodies. The slightest exertion overtaxed their weak frames, and death seemed always close.

Hour after hour, they would stand listlessly on the seashore watching the endless expanse of water. Whatever little hope they nourished of seeing an approaching sail quickly gave way to despair when the sun fell, and the fumes from the sulphurous springs cast a sickly greenish glow over the surface of the waves and turned the sea and the land into a phosphorescent nightmare.

Sometimes they would lie motionless in the shelter of the rocks, talking sadly of their former happy days. Then Shunkan would speak savagely of Kiyomori and with bitter hate would swear oaths of revenge. But the two younger men felt little resentment against him. He had pounced swiftly upon them for what was an act of treachery, and they were only suffering their deserts. In these long and weary days of reflection, they had come to feel that in their youthful enthusiasm they had allowed themselves to be swayed perilously near treachery to the Emperor Goshirakawa, and this they deeply repented. And so they besought Shunkan to be patient and to have hope that one day deliverance would come. But Shunkan, far from being comforted, only found cause for a yet deeper grudge in their sanguine youth, and he would curse them with such violence that they were glad to get away from him and leave him to his ravings.

Most of all, Yasuyori grieved for his mother. She was old, and since his father had long since died, Yasuyori had been her only support and comfort. He had been forced to leave without seeing her, and this was a constant source of sorrow and regret to him.

One evening, as he was idly carving on a piece of driftwood with a sharp shell, he conceived the idea of writing a poem to her and casting it on the waters. The idea soon became an abiding passion, and day after day he sought for strips of wood on which to carve the words which would describe his state to his grieving mother. When each was finished, he set it afloat on the tide. A day came when the hundredth frail vessel was launched on its haphazard journey, and he murmured a prayer that one at least might reach his mother's hands.

By chance, on the far distant shore of Itsukushima, a priest, who had just been officiating at the shrine of Akima, was meditating on a rock at the sea's edge, when an encroaching wave sent a small strip of driftwood dashing against his sandal. In some curiosity he picked it up and to his great astonishment found that it was a poem signed by Yasuyori, whose father had been an old friend of his. The poem was addressed to his mother, and the priest, who knew of Yasuyori's banishment, quickly concluded that Yasuyori had set the poem adrift in the hope that somehow it would reach her.

The priest lost no time in setting out for Mikoto where Yasuyori's mother lived. On arriving, he handed her the

piece of wood and told her of the miracle of its discovery.
When she read the poem and learned from it of the unbearable life on the island of their exile, she was deeply
disturbed. She put the poem on the straw matting of the
floor in front of her, where it lay, a silent message of
suffering and hardship, and her tears fell and drenched
her sleeves. The priest, no less moved, remained silent.
He watched the old lady in her grief and decided that he
would take the poem to the Emperor, tell him the story
of its discovery and, in humility, plead for Yasuyori's
forgiveness.

When the old lady's tears subsided, he told her of his
intention. She bowed in deep gratitude, and wrapping
the precious poem carefully in a silk handkerchief,
handed it to him. With her tearful blessings in his ear,
the priest started on his journey.

On his arrival at the palace, the priest was received
in audience. When the Emperor heard the story of the
piece of wood and read the poem for himself, he was
profoundly touched and perturbed. Seeing his concern,
the priest pressed his advantage and spoke most feelingly
on behalf of Yasuyori's mother, ending with a fervent
plea that the young man might be pardoned. The
Emperor said that he would consider the request favour-
ably, but added that the Lord Kiyomori must also be
consulted.

The next morning the Emperor called Kiyomori, his
son Shigemori, and the priest to wait on him. He
showed Kiyomori and Shigemori the piece of driftwood
and once more the priest described the strange story.
Shigemori was inexpressibly affected by what he had
heard, and even the uncompromising Kiyomori was
visibly touched.

'Your August Majesty,' began Kiyomori unexpectedly,
'although Naritsune and Yasuyori committed the grave

crime of treachery, they were never lacking in valour or humility. If they found themselves in this conspiracy, it was, I am willing to believe, due more to the influence of Narichika and Shunkan than to their own desire to be traitors. Both are still young; both have undoubtedly suffered unenviable hardships in exile and, as is evident from this poem, are repentant of their actions. They have learned their lesson and there is no point in wasting two promising young lives. I would ask your divine Majesty, then, to reconsider their sentence and to recall them to the capital.'

Then Kiyomori's face darkened and his brows drew together as he continued:

'As for Shunkan—that dastardly priest—let him rot there. I will have no hand in his pardon. He, with Narichika, were the principals in this revolt; if he returns he will bring nothing but trouble to us all.'

'Father, we all feel as you do,' said Shigemori, 'and I join my plea to yours that Naritsune and Yasuyori may be pardoned. I have it in my heart to wish that Shunkan, too, might share in your gracious Majesty's consideration of clemency, but I am well aware of the danger if Shunkan is released and free to wreak revenge on those who were responsible for his exile. His is a dark and scheming nature, and were he to return now with the thought of his present punishment to add to his bitterness, he would certainly spread revolt and treachery among those of the priests who are only too eager to change their priests' robes for armour and their praying beads for swords.'

Goshirakawa readily agreed to end Naritsune's and Yasuyori's exile and to recall them to the capital. Greatly rejoicing, the priest hurried back to tell Yasuyori's mother the glad news. Meanwhile a boat was prepared and a messenger sent on the long journey to the island.

It was almost three years that the exiles had languished there, and though the two young men had never given up hope of deliverance, they were almost at the limit of their endurance. Nevertheless, they kept up their daily vigilance, believing that while they could hold on to life, they could hold on to hope. But Shunkan cursed them for idiots and fools to trust in pardon as long as Kiyomori lived, and he turned his back on them and retreated to his rocky shelter, where he would sit for hours gazing fiercely before him as at some hideous inner vision.

One morning Naritsune, watching the horizon as he did each day, was startled by a sudden flash of white. Breathlessly he watched. Again! Something white! An illusion? A cloud? Or . . . a sail? Clutching Yasuyori's arm, he pointed:

'What is it, Yasuyori? Do you see it? Or is it yet another mirage?'

Together they stared, their hearts pounding and their limbs trembling. For what seemed an eternity, the fugitive white flash dipped and swung on the horizon, yet growing larger with every moment. Suddenly, with a cry, they stumbled to the water's edge. It was a sail and heading for the island! Almost bereft of feeling, they watched as it came towards them and anchored some distance out. They saw the figure of a man climb down into a small boat manned by a rower, who brought it speedily to the shore. Leaping out, the man waded through the shallows and stopped before them. For a long, tense moment they stared at each other—he, the elegantly royal messenger, and they, the wrecked and wasted exiles. Then the messenger spoke:

'Are you Naritsune, the son of Narichika? And are you Yasuyori, the former official?'

Taking a folded paper from his bosom, he continued:

'I am the bearer of a free pardon from His Most Gracious Majesty for both these persons. If you are the two mentioned, kindly answer to your names.'

They were weak from years of privation, and now, overcome with emotion, they could hardly gather enough strength to answer him. But they knelt and gave a ceremonial bow as they replied that they were indeed Naritsune, son of Narichika, and Yasuyori, the former official. The messenger handed them the pardon and they again bowed, until their heads touched the sands, but their tears fell fast and the words on the royal paper danced crazily before their eyes when they tried to read them.

Suddenly a wild, unkempt figure came running from behind a rock. It was Shunkan. Laughing madly and pushing and clawing at the others, he clutched at the paper.

'What! Has the high and mighty Kiyomori sent you to behead us? Does he still fear me so much that he must hound us even in our exile? But I won't die! I will live to revenge myself on him and all his house!' cried Shunkan.

The messenger looked sternly at him as he replied:

'Of that I know nothing. I come solely with an order for the release of Naritsune and Yasuyori.'

Shunkan stared at him stupefied; then dementedly he thrust his face forward to peer at the paper.

'My name! My name!' he yelled. 'Where is it? It must be there. The Emperor would not forget me. You are lying.'

The messenger thrust him back and motioned the other two to the small boat. Shunkan turned and caught at their tattered sleeves, piteously beseeching them not to leave him. They, as broken in body as he, begged the messenger to take him with them. But he refused, saying that the pardon was for two only and he must obey the order. Frantically Shunkan followed them, running now in front to push them back, now at their sides to hold and clutch them. Naritsune and Yasuyori in tormented voices begged him to be patient and they would surely persuade the Emperor to pardon him. At last they struggled free and were in the boat. The boatman pushed it out into the shallows, but Shunkan frantically threw himself across the gunwale. The small craft rocked and swayed as the boatman sought to free it from the old man's demented grip, and Naritsune and Yasuyori watched in helpless anguish as he beat and prodded savagely at the priest. Inch by inch he was forced back into the water, until only his torn fingers held in a last frantic and sliding grip. The boat sped on and the waves swamped him, but he seemed insensible to any but this last, desperate contact. The oarsman smashed at his fingers with the oar; he gave way, spent and defeated, and the onrushing tide hurled him back on to the shore. He raised himself, bruised and bleeding, and sobbed in a frenzy of despair, as he saw the others hauled aboard the ship and sail away into the distance.

Long he crouched there, dazed with exhaustion and abject misery, and oblivious to all but his consuming rage against Kiyomori. His fingers clutched at the sands as though they were clawing the life from his enemy's throat, and the saliva trickled from the corners of his mouth as he rained down curses upon the heads of Kiyomori and the whole Heike clan. When night came, he stirred and painfully dragged himself to a rock overlooking the sea. Demented, he filled the darkness with wild wailings, until, too weak to make another sound, he fell into a stupor and a temporary oblivion of the dreadful period of solitary exile that lay before him.

III

The Battle of Ichi-no-Tani

Some time after Munemori had succeeded his father as leader of the Heike, he learned that the armies of the Genji were advancing against him from the north and the east. One by one the provinces were rising up to join them under the Genji leaders Yoshitsune and his cousin Yoshinaka, and Munemori realized that any attempt to defend the capital would only lead to defeat and useless slaughter. After consulting with his captains, he decided to withdraw to the palace at Fukuwara near the valley of Ichi-no-Tani. Within a few days the withdrawal was complete, and shortly after the capital was occupied by the armies of Yoshinaka.

Munemori had deliberately chosen his father's old palace as the place for the Heike to recover their strength and spirit. It was situated in a district friendly towards the clan, and was a massive fortress. Built on a low ledge against the towering side of a mountain, whose base dropped perpendicularly to a narrow plain which ran to the sea, it was protected from behind by the savageness of the mountain terrain and the difficulty of descent from above. At the front it had an uninterrupted view across the narrow plain to the sea and any attackers from this direction would be in full view of the look-out towers. At some distance on each side, small hills jutted out from the mountainside and afforded some protection against any onslaught by horsemen. Here, thought Munemori, we shall be secure; for if the enemy gives battle, it will certainly not be at Ichi-no-Tani.

Taking advantage of a bloody feud which broke out among the ranks of the Genji, and which resulted in the crushing of the Yoshinaka faction by Yoshitsune, Munemori turned the palace into a fortified stronghold. Large contingents of warriors from the south reinforced those already at Fukuwara, and a new fighting spirit surged through the men. Calling his captains together, Munemori addressed them:

'Men of the Heike! The time has come for action. Our army is now strong and full of the will to victory. Our warriors are fired with the spirit of battle and determined to restore the Heike's glorious name to its rightful place in the capital. The enemy is weakened by internal feud and our fortress is impregnable. We shall not wait for the enemy to attack: we shall take the offensive. I therefore order you to prepare to march on the capital. Go at once and make ready; then return to me and we shall lay our final plans.'

But in Yoshitsune, the Genji had a leader who did not linger. After crushing his cousin Yoshinaka at the River Uji, he at once urged on his men to rout out the army of his enemy. Marching by night, and burning houses and farms to light the road for his warriors, Yoshitsune and his great army suddenly appeared before the Heike's fortress. Taken by surprise, Munemori hurriedly called together his captains and said:

'The enemy are ranged before us. They will need rest before they attack, and every minute of this respite must be used to deploy our warriors in battle formation. We are secure from attack from behind. We can therefore concentrate troops within the fortress against a frontal attack, and along its flanks to the left as far as the river banks, and to the right as far as the valley of Ichi-no-Tani. We shall anchor our ships off-shore across the bay. From them, our archers can harass the enemy who attack across the plain and provide a cover for our own men when they go to the attack. Warriors of the Heike, we must fight with courage and determination to annihilate the enemy, and enthrone the glorious name of the Heike once more in the capital!'

The captains withdrew, and that night they deployed the Heike armies in battle formation according to Munemori's plan. And none too soon; for next day

Yoshitsune launched his assault, concentrating his forces on the fortress and establishing his base in the valley of Ichi-no-Tani. All that day he attacked, but his soldiers were hurled back remorselessly with heavy losses, and some of his best commanders were left dead on the earthworks where they fell. By evening the assault had petered out, and the Genji armies were ordered to withdraw. Yoshitsune at once sent for his close friend and renowned captain, Benkei. After ceremonies of greetings had been performed and the attending retainers dismissed, Benkei graciously accepted the pipe and the tobacco-box handed to him by his lord.

'My friend,' began Yoshitsune, 'our plan has failed. Our losses have been heavy and we have not breached the enemy's lines at any point. Their fortifications are too strong to be taken by frontal attack, and if we continue such tactics we shall be crushed. If, however, we can attack from behind, we could take the enemy completely by surprise and victory would be ours.'

Benkei took the pipe from his mouth and looked at Yoshitsune with amazed incredulity.

'What, my lord? Attack from behind?' he cried. 'But that is impossible. The slopes of the mountains are as perpendicular as your lordship's golden screen. Neither man nor beast can ever hope to get a foothold there. We should all be dead before we could reach a Heike soldier.'

Yoshitsune's face took on a familiar obstinate expression.

'I am well aware of the difficulties,' he replied, 'but nothing is impossible and nothing impregnable. However formidable it looks, however strong their fortress, it is for us to find a weak point in their defences. We have made attacks from the sea and failed. We have made attacks along the flanks and failed. We are therefore left with the remaining possibility—an approach over the mountains and an attack from behind. Our first consideration then is to find someone who knows the mountains well and who would be able to advise us on the approaches to the summit. Could such a person be found?'

'That, my lord, presents no difficulty,' replied Benkei. 'There are many hunters in these parts, and if there is a way by which our warriors can scale the mountain, they would certainly know it. With your permission, my lord, I will take my leave and search one out immediately.'

All that evening Benkei roamed the countryside searching for someone to guide them. At last he found a boy, the son of an old hunter who had passed all his life among the mountains. The boy knew the mountains as well as his father, and when it was explained what was required of him, he agreed to accompany Benkei.

Benkei brought him to Lord Yoshitsune. The boy, overcome at being in the presence of so distinguished a person, fixed his terrified eyes on the floor, and it was some time before he could be made to speak.

'Can a man descend the slopes behind Ichi-no-Tani?' asked Yoshitsune.

'It is too difficult, my lord,' stammered the boy.

'Can a horse descend them?' asked Benkei.

'I have never seen one,' replied the boy, 'but I have seen deer descend.'

'You say a deer can descend them,' cried Yoshitsune. 'If a deer can descend those slopes, so then can a horse.'

The boy was dismissed and Benkei remained closeted with Yoshitsune until their plan of attack was agreed upon. Half of the army was to be deployed at each side of the palace, ready to create a diversion at the front. Guided by the boy, the other half was to follow Yoshitsune, each warrior leading his horse as silently as possible across the mountains to the summit where the slopes dropped to the valley and the rear of the fortress. The men below were to launch a three-pronged attack: across the rocky flanks on foot, and across the plain from the sea to the earthworks which stretched from the valley to the fortress. It was to be accompanied by battle-cries from every man to create a tumult that would penetrate to every corner of the fortress. Once joined in combat Yoshitsune planned to lead his mounted warriors down the slopes and batter the enemy from the rear.

'It will strain every man and horse to the utmost,' cried Yoshitsune in a stern voice when he explained his plan of attack to his gathered warriors next day. 'But where a deer can descend, so then can the men and horses of the Genji.'

58

The men were told to get what rest they could and then to gather under their respective commanders and make their way in silence to their assigned positions under cover of darkness.

As night fell, Yoshitsune and his men rode behind the boy to the foot of the mountains. There they dismounted and each man bandaged the hooves of his horse with sacking and removed all trappings that might give rise to any noise. Stealthily, and without a word, they began the difficult ascent. Shortly after midnight they reached the summit without mishap and at once composed themselves to lie quietly by their horses. They were ordered not to speak or whisper lest the breeze should carry the sound down to the watchers below.

Yoshitsune stood at the edge of the screen-steep slopes, and with only the sound of breathing horses in his ear, he stared at the fortress far below and the defences which stretched for miles on either side. Occasional sounds of singing and music came up to him on the night air, and it was evident that the thoughts of the dwellers within were far from any likelihood of an impending attack on their stronghold. Turning at a slight rustle, Yoshitsune found a dim figure at his side. It was Benkei. Together they stood in silence, at times leaning forward to peer below. Their thoughts were lost in the perilous task before them and their warriors, and Yoshitsune murmured a prayer and summoned his will and heart to stiffen his courage to the utmost. Gradually the sky began to lighten and a faint glow on the horizon heralded the first ray of dawn. Swiftly Benkei moved among the men and in a matter of moments all were seated in their saddles and ready for the attack. They looked at each other as the minutes passed, and here and there a warrior stroked the neck of his trembling horse.

Rigid at the head of his men, Yoshitsune peered

anxiously down. The clouds on the horizon were tinged with red, and the dawn streamed silver across the surface of the sea. Behind him a horse stamped the ground and snorted through trembling nostrils. Suddenly he stiffened as he saw a dark band of shadow detach itself from the surrounding gloom of dawn and move from the flanks and the sea towards the fortifications. Piercing shrieks and cries and the clamour of battle rent the morning air, and echoed and re-echoed against the mountain slopes and through the valley. Lights blazed up in the fortress, and the listening men above could hear the shouts and cries of the Heike warriors as they rushed to ward off the frontal attack. Not a movement could be seen at the rear; nor was there a watchman to report the menace overhanging them. It was as Yoshitsune had calculated —the rear was left undefended.

In a voice which rang clear above the clamour coming from below, Yoshitsune cried:

'The time has come to strike. Follow me and imitate the way I go. Lie back in your saddles and grip tight your reins. Never let loose the head for a second or you are doomed. Raise your swords to the sky and put your trust in the divine will.'

He seized his horse's bridle and reined in the bit until the animal's neck was arched as tautly as a bowstring. It plunged wildly with its forefeet as he forced it over the edge of the cliff. Flinging himself back flat in the saddle against the horse's rump, he lashed it furiously forward and the snorting terrified animal slithered down the slope in a rush of earth and flying rock. In a wild stampede Benkei and the warriors followed him. The crazed whinnying of the animals and the thundering clatter of hooves were choked in the clouds of dust that billowed around men and horses in their mad descent. Many fell and were crushed as they rolled, but the attack went on and man and beast were bruised and bleeding as their onslaught swept them downwards.

Milling and churning, they at last reached level ground. Cries of dismay and warning rose from the fortress as the Heike found themselves surrounded by an army which seemed to have descended from the skies. Heike warriors were rushed from the front to meet the new danger from the rear, and under a thunderstorm of arrows, the Genji troops swept up and down the wooden palisades with their torches until the palace fortress was ablaze in a hundred places. The Heike warriors fought desperately to stem the avalanche of men, swords, and arrows. But it was already too late. The outer defences were pierced and fighting was taking place inside the palace grounds. Their ranks broken, the Heike began to flee in wild confusion towards the boats anchored off the shore. All discretion gone, they succumbed to blind panic, and those who reached and scrambled into the boats safely, in their desperation slashed with their

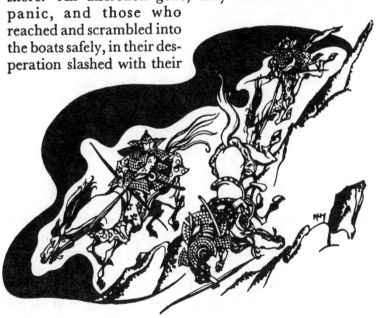

swords at the hands of those clinging to the sides. Thousands were drowned; the sea ran red with the blood of the slain; and the red banners of the Heike fell like maple leaves upon the waters.

Among the few who escaped the slaughter was Lord Munemori. He had under his care the child Emperor Antoku Tenno, and deeming it his sacred duty to put the safety of the child before all else, he escaped with the boy and members of his family early in the mêlée. News of his flight spread with the swiftness of lightning among the warriors of the Heike, and their morale utterly disintegrated. Those who had not escaped surrendered, and the few remaining pockets of resistance on the flanks collapsed. The victorious Genji were everywhere in hot pursuit of the Heike soldiers who had fled by land, and only the painful cries of the wounded and the dying broke the silence of the battle's aftermath.

It was at this moment of sudden quietness that a black horse came galloping from the burning castle to the shallow beach, bearing on its back a young warrior of the Heike. It was Atsumori, the sixteen-year-old nephew of Kiyomori. When the Heike lines broke, Atsumori had delayed to collect some of his small personal treasures, and turning to go, found himself confronted by three warriors of the enemy. A bitter struggle ensued, and although he got away safely, the delay proved fatal. When he reached the shore all the boats were gone and were already some distance off.

Atsumori gazed at the retreating boats in dismay. Knowing that all other roads of escape were blocked, he urged his horse into the water in the hope of swimming to the nearest boat. He had only gone a few yards, when a shout caused him to turn in his saddle, and he saw coming towards him a warrior of the Genji, mounted on a white charger and holding aloft in his right hand a black war fan with the bright red circle in its centre.

'What! Do I see only the back of one who, from his style of armour and manner, must be a noble of the Heike?' cried the warrior. 'Shame! Shame! Is it not our proud duty to show only our faces to the enemy? Turn and face me and we will answer to each other!'

The taunt of cowardice enraged the young Atsumori, and turning his horse he rode defiantly back to meet his opponent. Resolute and angry, Atsumori flung aside his bow, and unsheathing his sword, rushed violently on his enemy. The warrior drew his own sword and fought the boy off. Atsumori attacked again and again, but his sixteen years were no match for the strong veteran soldier, who in a swift attack flung Atsumori to the ground. The strict code of the fighting class now demanded that the warrior should behead his prisoner, and sheathing his fighting sword, he drew his short sword. Taking off the boy's helmet, so that he might cut off the head with greater ease, he discovered the handsome face of a youth beneath. As the warrior grasped the sword in both hands, he paused and his arms sank powerless to his sides.

'Alas!' he thought. 'So young and so brave! His looks and youth remind me of my own son, and how my heart would break were he to be captured and threatened with death as this boy is now.'

Dropping down on one knee, he placed his hand gently on Atsumori's shoulder and asked:

'Young sir, you are a noble of the Heike. Will you not honour me by telling me your name? I am General Kumagai.'

Atsumori looked at the face of this illustrious soldier. Tales of his brave deeds in battle and his benevolence of heart in peace were already legendary throughout the country. He was the pride of the Genji clan and among the Heike he was held in the greatest respect.

'I am Atsumori of the house of Taira and a nephew of General Kiyomori. I am not afraid to die. Fulfil your duty and behead me.'

Tearing his armour loose, Atsumori bared his neck and stiffened himself to receive the blow. But Kumagai turned aside and stood in sorrowful thought. This boy was of the noble Taira family and his parents, though aristocrats and his enemies, were but human and as other men and women. How desperate would they be in their grief when they heard of the humiliating death of their young son! He looked again at Atsumori, and touched to the heart by his innocence and youth, he determined to spare him and do all in his power to help him escape.

At that moment he heard the sound of hooves, and turning he saw in the distance a band of Genji soldiers approaching. Despairingly, and with tears filling his eyes, Kumagai turned again to his young captive and said:

'Honourable Atsumori, you are the scion of an ancient family. I wished to save you for the sake of your parents, and because you are young with so much of your life before you. But now I have no choice. My warriors are approaching; if I do not kill you, they will do so; and because I would not have fulfilled my duty, I too would be destroyed and the name of Kumagai would be for ever talked of in the same breath as cowardice. Therefore prepare yourself, young lord of the Heike. It is better that your death should be at the hands of one of your own rank than at those of a common soldier. I am sick at heart at all this senseless slaughter between our two clans. Forgive me for this deed! It tears my heart apart. Only know, noble Atsumori, that from this day on I shall cast away my sword and spend my days in prayer to atone for your death and for the comfort of your spirit.'

Tears fell from his eyes as he motioned the boy to prepare. Resolutely he lifted the sword in both hands and at one sharp blow severed Atsumori's head from his body. Wrapping the head tenderly in a cloth, he fastened it before him on the saddle, and brushing aside the questions of his soldiers, he rode away.

Kumagai laid aside all earthly pleasures, and after a period of meditation, submitted to the shaving of his head and became a priest. He lived the life of a hermit in the utmost poverty and humility and spent his days in prayer for the soul of the young Atsumori.

IV

The Fall of the Heike

After the disastrous defeat at the battle of Ichi-no-Tani, the Heike were filled with despair. Thousands of their army had been slain, and in the chaotic rout that followed, the warriors fortunate enough to escape the slaughter were widely scattered, many of them roaming the countryside without command or discipline. Munemori had succeeded in escaping with the child Emperor Antoku and the sacred regalia, together with some thousands of his retainers. Harassed for over a year by the armies of the Genji, he finally established himself in a friendly part of the country. Only too well he knew that his respite would be short. On sea and land, the Genji were gathering their forces to launch a new attack against him in an attempt to carry their recent success to a final and decisive victory. There was no time to be lost in rebuilding his shattered army. Scouts were dispatched to scour the mountains and plains for

the remnants of the Heike forces, with instructions for them to make their way at once to Munemori's temporary headquarters, where they were to be rested, reorganized, and fired with a new fighting spirit. Of his former great fleet of more than two thousand ships, barely five hundred remained, and these he brought together and amassed off the eastern coast.

When men and ships were assembled, Munemori took toll. Moodily he reviewed his ranks and his gloom turned to despair when he realized the paucity of their strength. But hope surged in him when it was reported that many units of the army had been seen in Kyushu and that a force of over a hundred ships under the command of the High Priest of the Kumano Temple, who was an ally of the Heike, had been sighted in the southern region of the Inland Sea off the coast of Kyushu.

Kyushu lay far to the south-west of his present headquarters. If he could act quickly, move his ships and men to that region to unite with his forces there, it would give him the precious time he needed to establish a fortress and deploy his warriors along the coast before the Genji began their assault. Yet he was full of foreboding. The next battle would be one of survival for his great clan. The Genji would attempt annihilation. All he could pray for was that, if the Heike could not turn the battle in their favour, they would fight till their last warrior had been slain, so that the glory and magnificence of their end would be told and sung by poets in time to come.

Before giving the order to carry his project into effect, Munemori turned his thoughts to the child Emperor. He needed someone to care for the child's sacred person and he bethought him of his father's former wife, the Lady Nii. At the time that Kiyomori retired to his solitary life as a monk, she too, as was the custom, had renounced all earthly ties and had gone to live in seclusion as a nun. Kiyomori was now dead and little of the outer world came to disturb her peace. When Munemori's message came to her, and she learned the full extent of the peril that beset the Heike and the person of the young Emperor, she had no thought but to leave her temple and undertake the high responsibility asked of her.

On the morning of the decampment, everything was in the confusion of preparation, and everywhere the tense faces of the men showed their recognition of the fateful issues involved. Only the child Antoku was oblivious of the future. A ship had been set aside for him and the Lady Nii, and from there he watched with delight as officers and soldiers moved to and fro. When all was ready, the warriors lined the decks to await their final commands and exhortations from Munemori. His messengers came aboard each vessel, and unrolling a scroll, read aloud in a fervent voice.

'Men of the Heike! Today we sail westwards. It will be our aim to link up with our forces in Kyushu and to establish a fortress on a suitably protected part of the coast. The forces of the Genji, I have learned, are already on the move to occupy strategic points along the shores

67

of the Inland Sea. It will require all our strength and resources of seamanship to reach our goal, if we are to avoid being drawn into battle before we arrive at our destination. Once united with our forces in Kyushu, we shall await the enemy with calm and determination. It will be a fateful battle. On it will depend our survival or annihilation. My warriors, I call on you to clear the great name of the Heike from the opprobrium of defeat. If it is the will of heaven, we shall succeed . . . if not, there must be no surrender! Every man must yield his soul gladly to join in honour his noble ancestors, and none must yield to suffer the ignominy of Genji chains.'

The words tore at the deep clan loyalties of the men, and from their throats a roar burst forth. General Munemori had never before received such ardent support from his soldiers, for his reputation had frequently been sullied with reports of stubborn, ill-conceived decisions, and a coarse and doltish nature. But now, in this time of peril, his words fired their hearts and they swore annihilation to the Genji.

In the young Emperor's ship, the guardian nun and her waiting maids listened to the message and the cries of the men in reply. Each was conscious, no less than the warriors, of the supreme importance of the coming struggle. The waiting maids wept. The child turned towards the Lady Nii. She met his serene, but questioning gaze, and in that moment she knew that he too was aware, in his own childish way, of what was at stake. He did not speak, but continued to look at her with his bright, noble eyes until she could bear it no longer. Turning swiftly away, she went into an inner room and fell on her knees in passionate entreaty to Buddha to preserve the pure young life entrusted to her care.

Outside a signal sounded; the pennants and flags were

hauled aloft; and in all the bright panoply of war the ships moved out to sea. For many weeks they sailed, their course set for Kyushu. The sea was calm and the wind was fair, and they made swift headway without any opposition from the enemy. On a bright spring morning, they entered the waters before the narrow straits of Shimonoseki. It was here that Munemori decided to land and erect a fortress which would become the rallying point of his forces in Kyushu.

Day after day he tacked and sailed, but at no point could he find a place to land. To his utter dismay, he found that the forces of the Genji had amassed all along the coast. Wherever he headed—to the north, south, east, or west—the enemy waited, implacable and in battle order. The Heike had sailed into a trap; they were cut off and caught in the pincer jaws of the Genji. In an endeavour to rally and link up with his forces on the mainland, Munemori sent out scouts under cover of darkness, but the news they brought back was disastrous. The men he had hoped would join him, disorganized and lacking leadership, had become weary and disheartened by constant defeat. One after another, detachments had gone over to the enemy camp to throw in their lot with the Genji. General Yoshitsune had lost no time in organizing them into his own divisions, and together with his own forces, they formed an impregnable barrier to the advance of the now depleted Heike army.

Munemori ordered his fleet to drop anchor at a point where he could observe the Genji ships and watch their movements. All that day and during the following night the fleet of the Genji, now a thousand strong, and that of the Heike turned idly on the waves of the flowing tide. When morning came the ships of the High Priest of the Kumano Temple rowed into the narrow straits, led by one bearing a standard inscribed with the name of the guardian god of the temple. To the dismay of Mune-mori and the watching Heike warriors, they rowed past their former allies and joined forces with the Genji. Despondency now settled on every warrior of the Heike. With their forces depleted, death and defeat stared them in the face. Only a miracle could give them victory.

Sensing the despair that swept over the whole fleet, Tomomori, the younger brother of Munemori, sprang into a small boat and rowed among the ships. In ringing tones that echoed among the decks above him, he shouted:

'Men of the Heike! The supreme moment has come. All of you know the seriousness of our position. There is no retreat. The enemy surrounds us on every side. But shall we idly wait to be annihilated by our hated foes? Shall we let ourselves become the laughing stock of the Genji? There is only one road for us. We must take the initiative. We must attack—and attack with all our strength and will to win. Our main aim must be to capture Yoshitsune. With Yoshitsune killed or in our hands, we shall not have won all: but we shall have dealt a moral blow to our enemies and that is half-way to victory. Let us then strain every nerve and sinew to seize him when we join in battle. If we succeed only in this, we shall meet our ancestors with unsullied glory.'

Tomomori's words dispelled the atmosphere of gloom from

the Heike ships. The warriors responded to them with the exultation of intense loyalty. The great fighting spirit of the Heike glowed in the eyes of every man. Whatever happened in the coming struggle, these warriors would cover themselves with glory. In the midst of a fervent display of devotion, young Noritsune of the Taira family leapt on to the gunwale of his ship, and looking fiercely about him, cried:

'Yoshitsune is famous for his leaping prowess, and well has he been named "the bird". Here stand I ready to leap into Eternity for the honour of the Heike. But first I will seize "the bird"; then shall we see if he can fly when I leap into the waves with him flapping under my armpit.'

Shouts of laughter greeted Noritsune's sally and the mood of the men was now keyed up in readiness to meet their fate.

The tide was running high. The signal went out for every man to stand prepared. It was Munemori's intention to raise anchor as the tide was on the turn, sail to Dan no Ura at the entrance where the waters were narrowest, and space his ships in tight battle formation. By this strategy, he hoped to reduce the area of action of the Genji ships.

Meanwhile Yoshitsune was not idle nor was he unaware of the great danger that lay in the fervent, implacable spirit that possessed the warriors of the Heike. There was little time to be lost, and swiftly he decided on his plan of campaign. Just as the tide began to turn and recede, he sent out a fleet of his small boats from their cover in the straits to where the waters ran fastest. Aided by the fast tidal flow, they were among the Heike ships almost before their anchors were safely bestowed. Like minnows darting among rocks the small boats flashed in and out among the enemy's vessels. All the time a deadly rain of arrows came from them, spreading panic and pandemonium among the surprised warriors on the Heike decks.

Tomomori and Noritsune rushed here and there, rallying the men, commanding and exhorting them, and soon the Heike were returning the attack with a pitiless hail of arrows. The air was filled with their screeching whine and the shrieks of those transfixed by them. Many hurled themselves from the Heike boats and boarded the little crafts of the Genji, turning their decks to rivers of blood with their swords.

The first fierce attack of the Genji was repulsed and the hopes of the Heike warriors soared high. It was at this moment that a pair of doves flew out of the blue sky and settled on the stern of Yoshitsune's ship. They were followed by a dark cloud which obliterated the sun and through which appeared the white flag of the Genji. The warriors of Munemori cried out in terror, and Shige-yoshi, a commander of some fifty ships of the Heike, reading these as ill-omens, deserted and crossed over to the enemy with the ships under his command. The Heike were now hopelessly outnumbered and Yoshitsune, guided by the traitor's information, marshalled all his forces and came back to the attack.

Backwards and forwards the battle raged, until the dead reached such dreadful proportions that hardly a vessel could be manned. The ships turned and swung on the tide, while everywhere on the encircling shores the armies of the Genji stood in readiness for the final assault. It was only a matter of time before the Heike must capitulate or choose the nobler path of suicide. Tomomori himself rowed to the vessel of the young Emperor, and calling the waiting maids together, he addressed them with sad dignified words:

'The end is near. There is no more to be done. You

must now address yourselves to Heaven. Take up all that is unclean or soiled and cast it into the waters. Cleanse and purify this vessel so that it may be a fitting threshold to the everlasting world. As true daughters of the Heike you will do what becomes you nobly and courageously. Farewell.'

From the waiting maids, he went slowly down to the stateroom where the Lady Nii sat watching her young charge. Tomomori bowed low as he entered the room, and leaning towards the nun, he spoke in a low voice. Her expression remained unchanged; not a muscle moved to show the import of his words. As Tomomori finished whispering, his voice broke and a look of sorrowful anguish fell across his face; but immediately he stiffened, and with respectful dignity and formality, bowed low to the Emperor and the nun in farewell.

The Lady Nii sat in silence after Tomomori left. Then smilingly she bade the child follow her into the inner room, where she helped him into ceremonial robes, while she dressed in the dark robes of mourning. Taking his hand she led him up to the upper deck. The waiting ladies were already gathered, dressed in formal kimonos. Everywhere the ship was swept and garlanded as for a festival. Lifting the child into her arms the nun bade the maids bind him securely to her body with her long sash. When he was securely fastened she took the two emblems of the crown—the sacred sword and the sacred jewel—and walked to the ship's side. Aided by her weeping maids, she stepped on to the bulwark.

The child looked up at her and asked:

'Are we going on a journey?'

'Yes, Your Majesty, we are going on a journey,' she answered, with the tears flowing down her cheeks. 'A long journey away from our sad and unhappy Japan to a land beyond the sea of everlasting joy and peace.

73

There you shall reign with all the virtues of your noble rank. But first we shall turn to the east where stands the Great Shrine of Ise and pray to the Guardian Deity for protection on our journey and say a last farewell; then to the west and offer up our prayers to the Lord Buddha that he may be there to welcome us to the country of everlasting bliss.'

They turned to the east and then to the west and prayed in silence. An inner serenity suffused the Lady Nii and a beatitude illuminated the beautiful face of the child Emperor. With her arms embracing him, she stood in her last brief moments a transcendent saint. In the next instant the swirling waters enfolded her and her sacred burden.

Transfixed, the attending maids stared at the circles of waves receding from the point where bubbles and foam were thrown up from the sinking bodies, until the last wave died and the bubbles ceased. Then one by one they offered up a prayer, climbed the bulwark, and hurled themselves to join their royal master and their noble mistress in the waters' depths.

From a neighbouring ship, the fiery Noritsune watched their heroic death. Tears streamed down his cheeks and fell on to the breastplate of his armour. With his arms hanging listlessly at his side, he turned away from the unbearable scene and gazed at the bodies of his comrades strewn in the contortions of death all around him. His sorrow suddenly gave way to violent rage and hate. He leaped into a small rowing boat and pulled with all his strength towards the ship bearing Yoshitsune's banner.

The Heike were doomed, but the battle was being fought with all the ferociousness of men determined to exact, before they died, the full penalty from the enemy. Arrows filled the air like rain; ships turned and manœuvred to close in violent death struggles; and the screams of the dying echoed along the shores. Oblivious to all danger and angered to a frenzy, Noritsune hurled his little boat through the churning waters and rode to the side of the ship of his greatest enemy. Yoshitsune watched him coming and stiffened, ready to meet him. His warriors rushed to form a barrier between him and Noritsune, who had now swung himself up the side of the ship and had reached the deck. The frenzied Noritsune rushed upon them with such an onslaught that they fell back, and he stood face to face with Yoshitsune.

'Defend yourself, for I have come to avenge my people. I am Noritsune, nephew of the great Kiyomori. In his name and in the name of the Heike, I challenge you,' he cried.

The soldiers flung themselves before him, but with a strength doubled by his fury and uncontrollable anger, he broke through their barrier and, laying fiercely about him, thrust them back, seizing first one and then another and hurling them into the sea. On they came again and again, but nothing could withstand the ferocity of his deadly sword blows. Once more he broke through their line, but Yoshitsune, seeing a group of Genji boats ride near and measuring the distance with a quick look, rushed past his enemy and sprang to land safely on the foremost boat. He turned and laughed in derision at Noritsune. It was a miraculous leap and Noritsune, despite his rage and mortification at his enemy's escape, inwardly admired the prodigious achievement.

Now the warriors resumed their attack and Noritsune was forced against the bulwark. But nothing they could do, no tricks of swordsmanship nor mass assaults, could cut him down. He fought back like one possessed. One by one they tumbled to his sword, until only a single warrior stood before him. Noritsune threw down his sword and rushed upon him. Seizing him with both arms, he lifted him bodily and stepped on to the bulwark.

'There is nothing left for me but death,' he cried. 'Yoshitsune, alas, has escaped me and I have to be content with you, my fine fellow, to accompany me to Eternity.'

Noritsune swung him under his arm, and with a last prayer to Buddha to cleanse and forgive him, he leapt with his captive into the churning sea where the engulfing waters closed over their heavily armoured bodies.

Yoshitsune, now reinforced by forces of his larger ships, closed in and overwhelmed the Heike flagship. All was now finished but the annihilation of the outnumbered Heike warriors, who were put to the sword without mercy. The sea ran red with blood and the smoke from the burning ships rose in clouds to blacken the blue sky. The warriors who escaped clung to the wreckage of their ships and were borne shorewards in the swift current of the narrow straits, only to be met and butchered by the waiting Genji horsemen who, throughout the great sea battle, had watched from the beaches at Dan no Ura. From the last remaining Heike ships came the sound of prayers as one after the other Tomomori and his comrades plunged into the foaming tide. Munemori, the last to leap, was seized with terror. In this dreadful moment when death was demanded as the supreme act of loyalty to the Heike, Munemori's weak and vacillating nature took hold of him. Scarcely aware of what he did, he floundered and swam desperately. Clutching at an oar which projected from a Genji ship, he clung frantically to it, oblivious to all but the terror of death. Hauled aboard, he was taken captive and some days later ignominiously beheaded in a deserted road on the way to the capital.

Thus the powerful clan of the Heike came by its downfall and was wiped out. It happened within four years of the death of its greatest leader Kiyomori, under whose sway it had risen to its glory. Marred only by the last cowardice of one man, it was an end full of heroism, sacrifice, and sorrow that will be told in tales and sung by poets for ever in the land of the children of the gods.

FOLK AND FAIRY TALES

The Peach Boy

By a river that flowed through a mountainous province of Japan, there lived a childless woodcutter and his wife. To all the children in the neighbourhood they were known with warm affection as Grandmama and Grandpapa, because, in spite of their poverty, they always remembered to keep a part of their small meal for the hungry young people who came to greet them daily. It grieved them greatly that they were childless, and always as their young friends turned homewards and their gay voices showered the evening air with their 'thanks' and 'goodbyes' and 'good nights', they would slide the paper-screen doors of their small house together and pause, silent and dejected, as if they could not bear to enclose the emptiness within.

One fine summer morning Grandmama decided to wash their winter kimonos in the river. In customary fashion she drew out the threads that held the long seams together and took the kimonos apart piece by piece. Gathering them into a rice-straw basket, she went down to the river bank. The river tripped and lilted in the sunshine, swinging round bends, sidling by rocks, and shooting through the mosses of the shallows. Skimming birds played with the flies, the flies with the jumping fish, and the willow trees dipped their long branches to the cool water. The river curled and eddied through Grandmama's fingers and round her sunbaked arms as she chanted 'Jabujabu, jabujabu,' to the rhythm of her washing. The sun and the river cheered her old wintry body, flooding it with the warmth of summer, and soon her work was done.

She laid the kimono strips along the bank to dry and was about to start for home, when she noticed a round object riding and tossing on the breast of the river, as it rounded the bend just beyond. It was buoyant as a cork and even the most ebullient eddies could not douse it. As it came nearer, she saw first the soft curved ripeness and then the bloom-furred skin of a golden peach. Its glowing skin, tinged with blushes, outshone the golden day. In all her life she had never seen a peach so big and so beautiful.

Quickly tucking her kimono above her knees, she stepped into the river and waded out until the waters brushed her gown. The peach rode jauntily towards her. She reached out her arms, but a sudden eddy sent it just beyond her reach, and there it spiralled and danced as if teasing her eager longing. She turned and started wading back, when suddenly the eddy moved towards her bringing the peach bobbing to her side. Folding her hands round the wet velvety skin, she lifted it out of the water and returned to the bank. She arrived home panting and breathless with the weight of it, but brimming with happiness at the thought of her prize brought safely home.

The dusk came and with it the clack-clock of her husband's wooden shoes on the stony path outside. Barely

had he time to put down his day's gathering of wood,
when Grandmama rushed out and flooded him with the
tale of her golden find. Grandpapa laughed at her
excitement as he kicked off his wooden shoes and stepped
on to the straw matting of their only room. Before him
lay the peach, its smooth and glowing warmness filling
the whole room. His eyes, still closed in a twinkling line
of laughter, slowly opened in amazement. He touched
the peach to assure himself that it was real and then
sat down on the floor, his legs tucked under him, to gaze
at this miracle of all peaches. It would be a glory to
eat!

When they had finished their simple meal of rice and
dried fish, Grandmama cleaned the big kitchen knife.
Together they put their hands on the haft and gently
sliced down the golden cleft of the peach. As it fell apart,
there was a stirring in the heart of the fruit. They fell
back in fear as a boy, gay as the first green of spring,
stepped out before their astounded eyes. He smiled at
their incredulous wonder, and with a swift confiding
movement, turned to Grandmama and pulled the folds
of her apron around him. He rested tranquilly against
her knees while the old couple sat spellbound. Long
they remained, silent and motionless, but with hope
slowly rising in their hearts that Buddha had at last
relented and sent a child to them in the evening of their
days.

And so indeed it proved. For as the seasons of the
planting and the gathering of the rice came and went,
the boy gave nothing but pleasure and happiness in his
new home, and Grandmama and Grandpapa never
ceased to rejoice in their good fortune. They called him
'Momotaro'—the son of a peach—and as the months
rounded into years, Momotaro grew sturdy of body and
firm of limb. His skin glowed tan and rosy-gold and he

carried with him a stoutness of heart and a sweetness of disposition that might well be attributed to his strange fostermother.

One day, soon after he had reached his fifteenth birthday, Momotaro asked leave to address his parents on a matter of great importance. Greatly wondering, they waited for him to speak. Momotaro bowed low to them in filial piety, then said:

'Honourable Grandpapa! Honourable Grandmama! Though I became your son in a most unusual way, I can never cease to be grateful to you for the good but disciplined manner in which you have brought me to manhood. Your kindness has been wider than the horizon of the sky and your love has flowed over me with the fulness of the river that brought me to you.'

Never before had they heard him speak with such seriousness. No longer was it the voice of their gay child. Before them stood a Momotaro, for all his smallness and youth, grown into manhood. Overcome with the tenderness of Momotaro's words, and at the same time sensing that behind their seriousness lay some firm decision that would affect the joy of their home, Grandmama began to weep. Grandpapa, remembering the days of his own youth and the first time he had wished to prove his manhood, guessed that Momotaro was longing to go into the world to try his fortune, and before the boy could continue he said:

'My son, I can guess what you are going to say. And though we shall taste the bitterness of loneliness without you, yet I applaud the courage of your spirit that prompts you to this course. Please do not allow our sorrow or the tears of your foolish mother to deter you. Only, we beg you, always remember that we shall be waiting for you as long as our fading years permit, and our poor hut will for ever be your home.'

'You read my thoughts well, honourable Grandpapa,' replied Momotaro with the same serious bearing. 'And when the time comes for me to leave, your understanding will make our parting easier.'

'But when will you go?' wept Grandmama, unable to control her tears. 'When will you go? Beyond the hills of our village and the river of our valley, the world is angry and evil. It is no place for you.'

'It is beyond the hills of our village and the river of our valley that I must go, and without a moment to lose,' said Momotaro. 'And for no other reason than to quell that anger and to put good where evil reigns.'

He paused and then continued:

'It is a long story and I fear that I shall weary you with the telling of it. In the ocean that washes the shores of our country, there is an island of evil. It is inhabited by fearful horned ogres. They are taller than the tallest bamboo in the forest and their hearts know only darkness. The skins of some are red as the belching flames of Mount Fuji, some blue as the ocean's stormy depths, some green as the eyes of cats at night, and some are as black as the evil in their hearts. They come in swift boats to our shores to ravage and pillage the countryside, devouring the children and leaving grief, destruction, and death everywhere behind them. I am going to brave these monsters on their island and exterminate them. I will bring back to Japan all the treasures they have stolen and restore them to their rightful owners.'

At this Grandmama's tears redoubled, but Grandpapa checked her and said sharply:

'Do not be so foolish, wife! This is no time for tears and certainly no time to display them. Our son is brave. His cause is just, and Buddha, who sent him to us, will protect him. Until today he has been a boy: now he is to prove himself a man. Therefore, cease these useless

tears and help the boy to equip himself for the long journey and the battles that lie before him.'

With these words, Grandpapa went out to the woods to cut Momotaro a stout staff. Grandmama, drying her tears, took from the cupboard a bag of millet and began grinding the seed in a heavy mortar. Then she made the ground millet into dumplings and cooked them over the charcoal fire. Very soon Grandpapa returned with a stout branch. He stripped off the bark, baring the gleaming white wood which he speedily polished into a shining staff.

When all was prepared and Momotaro was about to leave, Grandpapa took from a lacquer box a warrior's iron fan and, with many exhortations and blessings, handed it to Momotaro, who tucked it into his kimono sash. All knelt on the floor and bowed deeply many times. No words were spoken and no grief showed in their faces. The depth of their silent bowing expressed only too well the sorrow in their hearts. Finally Momotaro took his leave, and his grandparents watched the valiant little figure stride into the distance. For a moment he turned, and all bowed their last farewell before he disappeared over the brow of the hill. Now they could only wait and offer prayers at the family altar for his safe and speedy return.

Momotaro, once he had overcome the sadness of

parting, strode forward full of exuberance. It was good to be out in the world and, come what might, he knew the justice of his cause. He walked the whole morning through and deep into the afternoon. The rice fields had long been left behind him as he climbed hill and mountain. Peak after peak had passed, and now a valley with a coppice in it lay before him. He stepped aside into the trees and for the first time since he left home, prepared to smooth the raw edge of his hunger. He had hardly untied his bag of millet dumplings when he heard a sudden scuffling behind him, and leaping to his feet he saw a large and lordly dog bound from an entanglement of bushes. The dog snarled fiercely and barked in a deep voice:

'Who gave you leave to travel through my country? I am the Lord Brindled Dog, and all who come here must obey and pay homage to me. If they don't, I bite off their heads!'

He snarled again and looked as fierce as a jungle full of tigers. Momotaro, instead of melting with fear, started laughing at his comical face.

'Lord Brindled Dog, I think you are more bark than bite! I am not afraid of you!'

With these words Momotaro waved his staff before the lordly creature, who suddenly, with his tail between his legs, retreated to a safe distance before saying in a most conciliatory voice:

'You must be the famous Momotaro—the Momotaro that even the winds speak of, and of whose coming the pike in the river tell tales. Do me the honour of informing me what brings you into my domain.'

Momotaro smiled at the Lord Brindled Dog's sudden change of tone, but willingly told him of his plans to exterminate the ogres. Lord Brindled Dog's ears and tail reared stiff in excitement.

'Those are the monsters who killed my brindled heir

and devastated my lands. I vowed vengeance and now my hour has come. Momotaro Sama, take me with you. I am big and strong. I can run faster than the fastest creature on land. With my crunching jaws and sword-edged teeth, I can snap their heads from their bodies in one bite.'

'The proof of that pudding will be in the eating of it,' chuckled Momotaro, and he agreed to have Lord Brindled Dog as his fighting companion.

They set out together, having first refreshed themselves by sharing one of the millet dumplings, and travelled quickly on the way. Lord Brindled Dog was ever speeding ahead, snuffling the ground and smelling out the best track for them to follow. Mile after mile sped under their feet, until, towards evening, they came to a small hollow among the foothills of a mountain. Here they decided to rest for the night. They settled down under the spreading branches of a tree, but hardly had their heads touched their pillows of leaves, when there was a rustling commotion overhead and a handsome monkey came swinging down with long-armed grace to their feet. He bowed to Momotaro and said with charming politeness:

'You are undoubtedly the Lord Momotaro of whom all the forests and mountains have heard. I have come to offer my services in the task of right and justice you have undertaken. I am the Lord Monkey of this Mountain and I beg you to accept me as your retainer.'

Lord Brindled Dog, when he heard these words, jumped snarling forward and yelped:

'I am Lord Momotaro's retainer and he needs no other. What use would a monkey be in making battle against monster giants? Be off with you to the tree-tops where you belong.'

But the Lord Monkey of the Mountain remained calm and looked steadily at Momotaro. Momotaro liked him at once and his mind was made up.

'You, Lord Brindled Dog, if you really wish to serve me faithfully, will remember that there is arduous and dangerous work ahead of us. You will need all your battle energy for that time. Do not waste it now. Come, Lord Monkey of the Mountain, here is a millet dumpling. Share it with Lord Brindled Dog. I am happy to have you as my second retainer.'

As the first hazy light of morning filtered through the branches above them, Momotaro and his two retainers rose up, yawned, and set off on their way. The air was full of the perfume of wild flowers, and the trees and bushes were astir with the movement and song of the little bush-warblers. It seemed a day full of promise to the three warriors as they strode on their way. Lord Brindled Dog bounded ahead with Momotaro's staff; Lord Monkey swung gracefully above their heads from tree to tree; Momotaro himself strode out blithely, carrying his fan in his hand as befitted a lordly warrior. Foothills gave place to forests, forests to winding streams, and winding streams to moorland. It was country that delighted Lord Brindled Dog's heart. He ran and leaped, and snuffled and panted. Suddenly before his quivering nose a pheasant rose out of the gorse. Startled, the dog dropped Momotaro's staff and immediately sprang forward to bite off the pheasant's head. The pheasant, undaunted, whirled in the air and swooped to attack the dog. Momotaro watched and thought, 'What a valiant creature! Just the follower I would have with me.'

He picked up the fallen staff and brought it down with a thud on Lord Brindled Dog's rump, sternly ordering him to stop fighting. At the same time he spoke sharply to the pheasant and said:

'Who are you that dare molest my personal retainer? We are bound on a most important expedition and you are causing us much delay.'

On hearing Momotaro's voice, the bird crouched low near his feet and said:

'You must be the great Lord Momotaro of whom all the moorfowl have heard. I am the humble Lord Pheasant of the Moor. I and my bird retainers know of your noble undertaking. We are whole-heartedly behind you and my only wish is to be allowed to serve under your leadership.'

Momotaro was delighted with his new ally, and taking out another millet dumpling, he shared it among them all. Then all four continued across the moorland, the monkey riding on the dog's back, and the pheasant flying overhead and occasionally perching on Momotaro's shoulder.

The morning passed to noon and soon the scrub-covered moors were far behind. A wood of tall bamboo trees stretched endlessly through the afternoon and the four valiant heroes walked, ambled, hopped and swung under the cool shade of the leaves. Dusk was just falling gently over the forest when they emerged to see before them the blue expanse of the ocean. All four sat down, dusty, tired, and hungry, but overjoyed to be facing their first objective. They scanned the sea's glassy surface, but nothing rose to break its quiet immensity.

'We will sleep here tonight and tomorrow morning we will build a boat and start our search for the ogres' island,' said Momotaro.

At the first light of dawn Lord Momotaro was up directing his three followers. Lord Brindled Dog felled the bamboo trees with savage bites of his fierce jaws; Lord Pheasant of the Moor brought long strands of creeper from distant fields to bind the trunks together; Lord Monkey of the Mountain and Momotaro worked quickly and deftly and soon their craft was shipshape and launched. Lord Brindled Dog took charge of the oar, which swivelled on a wooden pin at the stern of the boat. His strong, powerful, curving strokes sent the boat speeding through the waveless blue. The land fell out of sight and the sun rose high in the sky. All day the little boat moved farther and farther out to sea, but from one end of the horizon to the other there was no sign of an island to be seen: nothing but the vast murmuring expanse of the ocean.

'Lord Pheasant of the Moor, now is the time for you to use your special gift of flight,' said Momotaro. 'Go before us and see what there is to be seen from your vantage point in the sky.'

The pheasant rose swiftly in the air and flew towards the horizon, mounting steadily higher and higher until he vanished from sight. It seemed a long time to those waiting before the bright wings at last reappeared.

'Lord Momotaro, I have seen the Island's outline. It is still far away. But with good luck and swift rowing we should reach there before nightfall,' cried the pheasant overhead. 'Follow me and I will direct you.'

Following the pheasant who flew on ahead, Lord Brindled Dog rowed with renewed strength and sent the boat skimming over the water. It was not long before a dot appeared on the skyline, which gradually grew into the outline of the island they sought. As they drew closer, its grim, forbidding look, its air of iron-clad isolation, cast a gloom upon the whole surroundings and made even the peach-stone heart of Momotaro chill for a moment. Looking at his followers, he saw that the dog's hair bristled along his spine and the monkey was chattering to himself in quiet undertones. Only the pheasant, now settled on his shoulder, looked unconcerned, his beady eyes as bright and alert as ever.

At the command of Momotaro, Lord Brindled Dog rowed quietly inshore. From there Momotaro could just see, through the falling dusk, the ogres' fortifications: a high iron-railed fence on the lower reaches, and behind, but farther up the rocky hill, a heavily spiked wall which encircled a massive fortress-like building. 'That,' concluded Momotaro, 'holds our enemies.'

He directed the boat to ground in the shelter of a large rock which jutted out beyond the iron-railed fence, and there took counsel with his followers. He explained to them his plan of attack, which had been formulating in his mind since he first saw the island's defences. They would attack under cover of darkness, but first they would rest a little to regain their strength. It was their best chance of success, if shrewdness, which they must depend on, was to beat the mighty ogres.

At the chosen hour, they started for the iron fence. Lord Brindled Dog quickly made an opening by biting through several of the uprights, and he and Momotaro stationed themselves one at each side of the massive wooden doors in the inner wall. Lord Monkey scaled the wall and waited at the top. Lord Pheasant went winging high over the fortress and alighted on the topmost turret. Suddenly, in the still darkness, the pheasant shrilled out his clear challenge, 'Ken-ken, ken-ken, ken-ken,' a cry which he knew all mortals took for the warning of an earthquake. The ogres, bleary-eyed with sleep, stumbled out into the darkness of the yard. The pheasant screeched again, while the monkey ran along the wall, tearing up the stones and hurling them at the heads of the ogres. Bewildered, the monsters flung open the great wooden doors, their howls filling the night. As they came through in blind confusion, Momotaro struck their legs with his dagger and, as they stumbled, Lord Brindled Dog tore their heads from their bodies with one snarling snap of his iron jaws. Inside, Lord Pheasant swooped in fury out of the dark sky and blinded them with his sharp

beak; while Lord Monkey tore the hair from their heads with his strong fingers. The few that escaped rushed in panic to their boats, but were blinded by Lord Pheasant before they could reach them.

Only the ogre chief was left when morning came over the distant skyline. Surveying the ravages about him, he knew that defeat had come to him at last. He broke off his horns and taking the keys of his fortress treasure house, he laid them at the feet of Lord Momotaro in token of submission. Leaving Lord Brindled Dog to guard the ogre chief, Momotaro, with the pheasant and the monkey, went to ransack the storehouse for the treasures he had dreamed of returning to Japan. Many he saw that he thought had disappeared for ever: the precious stone with which the tides could be controlled;

the garments which rendered their wearer invisible; the mallet which struck showers of gold at every blow; the precious coral which a famed Empress brought from the depths of the ocean; and many priceless treasures of musk, emeralds, tortoiseshell, gold, silver and amber. All these they loaded into one of the ogres' boats and then set out for home, leaving the ogre chief solitary among his evil dead.

The flowing breeze bellied their sails: the tide ran fast in their course: and before long Lord Pheasant of the Moor announced from his heaven-high look-out that the coastline of their beloved Japan was in sight. Joyfully they landed and at once set about building a little cart, on which they loaded all the treasures from the ogres' storehouse. Soon they were eagerly striding homewards through the forests and moorlands.

As Momotaro descended the last hill with his faithful followers, there before him lay the village, with Grandpapa and Grandmama standing at the door of their little hut waiting to welcome him home. They bowed deeply and their eyes told the depth of the happiness in their hearts. The time had seemed endless while he had been away, but now it flew faster than the wings of morning as he told of all that had befallen him.

Momotaro's grandparents lived for many golden years to enjoy the good fortune their wonderful son had brought them, and Momotaro became a powerful but kindly and just lord as the years went by. Lord Brindled Dog, Lord Monkey of the Mountain, and Lord Pheasant of the Moor remained always his closest friends, paying him frequent visits from their own domains, where he in turn was always a most honoured guest. And Momotaro spent the rest of his life in using the powers that the ogres' magic treasures had given him for the welfare of his people.

The Old Man Who Made the Trees Bloom

ONE morning, long, long ago, an old woodcutter, who lived in a small hamlet by the side of a great forest, was on his daily journey to cut down trees for the lord of the province, when he noticed a little white dog lying by the side of the path. It was thin and wasted and near to death from cold and hunger. Moved to pity by the creature's suffering, he picked it up, put it tenderly into the breast of his kimono, and carried it back to his house. His wife hurried out to meet him when she saw him return so early, and asked what was the matter. In reply he uncovered the little dog and showed it to her.

'You poor little dog!' she cried in sympathy. 'Who has been so cruel to you? And how intelligent you look with your clear bright eyes and alert lively ears! An old couple like us might well like to have you in our house.'

'Indeed! Indeed!' murmured the old man, who was only too willing to have it as a pet. They both went inside, laid it on the straw-matted floor, and began at once to attend to its sickness.

Under their tender care the little dog grew well and strong. His bright eyes grew brighter, his ears stood erect at the slightest noise, his nose was forever twitching from side to side with eager curiosity, and his coat gleamed with whiteness, so much so indeed, that they called him 'Shiro', which means 'white'. As the old couple had no children of their own, Shiro became as dear to them as a son and he followed the old couple wherever they went.

One winter day the old man, his spade over his shoulder, went to his field to dig some vegetables. Shiro, always happy on such occasions, jumped and frolicked in great circles round him and made sudden raids into the ditch and undergrowth. When they reached the field, he careered as madly as ever and barked in delight as he hurled himself over the brambly bushes. Suddenly he stopped. His ears went stiff and erect and his whole body became alive and tense. With his nose to the ground he moved slowly to the fence near the corner of the field. His nose twitched and sniffed over a little mound of earth. All at once he began digging furiously, sending the earth through his hind legs in a continuous cascade. His loud excited barking attracted the old man's attention at the other end of the field, and thinking that Shiro must have discovered something very extraordinary to make him behave in such a way, he came hurriedly to see what it was all about. Taking his spade, he started digging in the hole scraped by Shiro, but hardly had he removed two clods when a shower of golden coins poured into the air as though from an invisible fountain. The old man fell back in wonderment and hurried away to bring his wife to see the miraculous sight.

Their neighbour, an ill-tempered and avaricious man, had also been attracted by Shiro's barking, and from the other side of the bamboo fence separating their fields he had witnessed this unbelievable wonder. His mean eyes glistened with covetousness and he could hardly control his grasping hands. Cunningly he put on a friendly voice and begged the old couple to lend him their dog for the day. Gentle and kindhearted and wishing to be of service, the old man lifted Shiro and, telling him to be a good dog, handed him over the fence to his neighbour. Sensing the man's ugly nature, Shiro refused to follow his temporary master. He crouched on the ground, his trembling body gathered up in fear. The neighbour coaxed and shouted, shouted and coaxed, but only succeeded in further increasing Shiro's fear. Growing more and more angry, he tied a straw rope round Shiro's neck and dragged him forcibly to a corner of his field where he bound him to a tree, so tightly and with such a short lead that the poor creature was forced to lie in an agonizing position. His throat was so constricted by the rope that his weak barks could not be heard by his own master.

'Now then,' shouted the vicious neighbour, 'where is it buried? Where is it buried? Seek it out for me or I'll kill you, you vile hound.'

Furiously he struck the ground before Shiro's nose. The blade slid into the earth and scraped against some metal object. The surly man stood still. His eyes widened in avid expectation. The next moment he was clawing the earth with both hands in a frenzy of greed. When he unearthed nothing but old rags, wooden clogs, and broken tiles, his fury became uncontrollable. He picked up his spade and struck viciously at Shiro, who was whining and cowering in terror at the foot of the tree. The blow cut him cruelly, but it also severed the rope that held him, and Shiro ran in anguished circles, dazed by the blow and howling pitifully. Attracted by his cries his master hurried to the fence, and when he saw what had happened was beside himself with grief. Shiro crawled through the fence and his master gathered him tenderly in his arms.

'Shiro, my poor Shiro, what terrible thing has happened to you? Will you ever forgive me? Will you ever forgive my cruel mistake?' the old man sobbed. But Shiro could only shiver and cling more tightly to him.

Sadly the old man returned home with his pet. He bathed and tended his wound and fed him with thin gruel. But despite all his efforts, the spade of their wicked neighbour had wounded him so severely that he died that same evening.

The old couple were overwhelmed by their loss. They did not sleep that night, and early next morning, with great sorrow and mourning bitterly, they buried their little pet in the corner of the field where Shiro's miracle had happened. Over his grave the old man constructed a small tombstone, and beside it he planted a young pine tree. Every day, the old couple went to the grave, and standing side by side with bowed heads, they mourned for their friend.

The tree grew with incredible swiftness. In a week's time, its branches shadowed Shiro's grave; in a fortnight's time it would take two people with their arms stretched round to span its trunk; and within a month its topmost leaves seemed to sweep the sky, so tall had it grown. Each day the old man, gazing upon this new wonder, said:

'Wife, truly another miracle. Our little Shiro is dead, but his spirit has entered into this tree. His gaiety and exuberance could not be killed. It has become the sap of this magnificent tree and is prancing madly in its leaves and branches. I am sure of it.' And they gazed upon the tree with renewed astonishment.

The news of the growing tree soon spread. From distant mountain and valley people came and gathered round it daily. They craned their necks and strained their eyes to see its topmost branches, now dimmed in the haze of the sky. They shook their heads and whispered to each other that it was not true, but raised their heads to look again and could not doubt their eyes.

One day in the winter the old woman said to her husband:

'Husband, do you remember how our little Shiro loved rice cake—mochi-cake? Would it not be a fine idea to fashion a good mortar from the trunk of Shiro's tree and make mochi-cake to offer on his tomb?'

'What a fine idea, indeed! What a fine idea!' her husband replied excitedly. 'We will certainly do as you say.' And immediately he began sharpening his large axe.

The following morning and all the afternoon he worked, and slowly his blade ate into the great trunk. At last, with one powerful final swing, the great tree creaked and fell to the earth with a roar so mighty that it was heard in the furthermost corners of Japan. A fine and beautiful mortar came to shape under the skilful hands of the old man, and was soon ready to receive the glistening white rice for pounding. With hearts full of love and tenderness for the memories of their little friend, the old couple began to beat the rice with their pestles to turn it into a fine powder before cooking. But hardly had they broken more than a rice-bowl full of the grains, when before their amazed eyes the whole pile turned into a glittering heap of golden coins.

How they marvelled! And how eagerly they talked about their good fortune to the neighbours, who were all overjoyed that such riches should fall to them. All, that is, except the mean irascible man who had so cruelly killed Shiro. He could hardly contain his greed as he listened to the story of the old man and schemed,

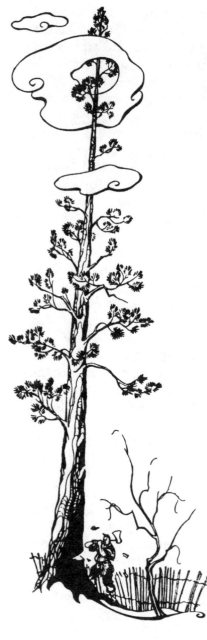

there and then, to gain possession of the magic mortar and have it for his own. Next day he went to the house of the old couple, and cringing and fawning and feigning great sorrow, said:

'Ever since the death of your little dog, I have been filled with a great melancholy. A great melancholy, good neighbours, because I feel I was to blame. Night and day I have thought that if only there was some way in which I could show how deeply I feel about it and do something to show my repentance, how gladly I would do it, but I was too ashamed. Today, in all humbleness, I have come to ask your forgiveness. I would so much like to make some fine mochi-cakes and offer them on the tomb of little Shiro. But, alas, my mortar is too old and despicable and I am too poor to buy a new one. Would you, kind neighbours, lend me yours for a little while to make my small offering for our little friend?'

The fond and foolish old couple were deeply moved

by this deceitful talk, and believing that he was sincerely repentant, they allowed the scheming rascal to take the mortar with him, and he gleefully rolled and tumbled it before him to his house.

On arriving home, he lost no time in preparing to make the cakes. Together with his equally avaricious wife, he poured the rice into the mortar and they set about pounding it. On and on they pounded but no gold appeared and they both shouted angrily:

'Turn into gold, you miserable grains! Turn into gold!' And they hammered and hammered more vigorously than ever. 'Don-don, don-don,' went their pestles, and the grains went flying in all directions, but not a single coin of gold flew out. They were about at the end of their strength, when suddenly the pounded rice in the mortar began to move and transform itself.

'It is changing,' cried the old knave.

'We shall be rich,' cried his shrewish wife.

And they danced with delight round and round the mortar. But to their horror, instead of a fine heap of glittering gold appearing, there came nothing but old rags, wooden clogs, and broken tiles, just like the rubbish he had dug up in the field. In a great rage he seized his hatchet and with one blow split the mortar in two. His wife seized another hatchet and in a frenzy they both hacked and chopped the two halves to pieces. Lighting a fire in the kiln oven, they flung the pieces in and watched until they were consumed to ashes.

Next day, the old man came to ask for the return of his mortar, but the neighbour gave him a surly greeting.

'The mortar was cracked and useless. At the first stroke of my pestle it fell apart, so I chopped it up for firewood and burnt it to ashes. If they are of any use to you, help yourself. There they are in the kiln.'

With these curt words the neighbour crabbedly turned his back and refused to say another word.

The old man was desolate. He looked first at his neighbour and then at the kiln. His eyes rested on the ashes. There was no anger in his heart, only a deep sadness.

'First my dear Shiro, now my wonderful new mortar,' he lamented to himself. 'A coldhearted and unfeeling man! Yet, what is to be done? Nothing, no, nothing can restore them to me. Only ashes left. But they are the ashes of my little dog; for truly the mortar was made from his divine and wonderful spirit. I will take them and bury them beside him. He will be overjoyed to know that his spirit has been restored to him.'

The old man gathered the ashes into a rice basket and turned slowly homewards, wondering what his wife would think of this new disaster. He had hardly travelled more than half-way, when out of a pine grove a gentle breeze arose, danced momentarily among the trees, and the next instant was swirling round the rice basket and lifting the ashes high into the air. The breeze died as quickly as it had risen and the ashes floated down like snowflakes on the cold naked branches of the wintry trees. But wonderful to behold, wherever they alighted, the naked branches burst into a profusion of blooms and leaves, and soon everywhere about the dazed old man, the cheerlessness of winter was transformed into the gaiety of spring and the air was filled with the perfume of opening blossoms. The old man turned slowly to gaze upon this new wonder. He stretched out his hand to touch the leaves and petals to assure himself of their reality. He turned slowly round and round, his eyes drinking in the young greenness and his nostrils filled with the fragrance of May. Suddenly he was running excitedly to the village.

'Look! Look! Old Flowerman can make the trees bloom! Old Flowerman can make the trees bloom! Look! Look!' he cried, throwing handfuls of the ash on every tree and bush, and watching how the trees and the bushes opened in bloom where it fell.

It happened that the lord of the province, accompanied by his retainers, was making a tour of the village. Attracted by the shouts of the old man and the crowd of people surrounding him, the lord reined in his horse and asked one of his followers to find out what the excitement was.

Meanwhile the old man, in his unbounded joy at the new beautiful power he possessed, had climbed a cherry-tree, and singing all the while to himself, scattered the ash on every branch, and the pink and white blossoms spread their radiance about him.

The retainer called to him, and descending from the tree, the old man was taken into the presence of his lord. Humbly and simply he told his story, and when he demonstrated the miracle of the ash, the lord was filled with great pleasure and said:

'Wonderful! Truly wonderful! A man who causes flowers to follow his shadow! Where is there another who possesses such a gift of beauty? Old man, I shall reward you,' and he dismounted from his horse.

A retainer brought an offering table and on it placed a rare brocade bag filled with golden coins. The lord himself held it out and the old man, bowing first to the ground, took the brocade bag with humble reverence.

He could hardly wait to get home to tell his wife of the miracle of the ash and the honour that had been done him by the lord of the province, and as he hurried along clasping the precious bag, he was filled with delight and smiling with pleasure.

The greedy neighbour was a witness to the whole happening and was filled with bitterness and resentment. He hurried back to his house and opened the kiln door. Sure enough, there were traces of the ash in the oven and on the floor. He called to his wife and they both scooped up what was left into a basket. With the basket under his arm he hurried out and waited by the roadside for the return of the lord and his train. The sound of the horses' hooves told him that the retinue was approaching. He quickly climbed the nearest tree and began singing to himself and calling out, 'Old Flowerman can make the trees blossom! Old Flowerman can make the trees blossom! Look! Look!' just as the old man had done.

The lord trotted his horse to the tree and, looking up, said:

'What! Have we another worker of miracles in the village? This is certainly not the same old man as I saw just now. You there, are you another who can make the trees bloom? If so, demonstrate your powers at once.'

'Yes, my lord, I will do so at once,' replied the ill-natured neighbour.

Straightaway he began sprinkling the ash over the branches. But instead of settling and bringing forth flowers, the ash flew helter-skelter in all directions and enveloped the lord and his retainers in a choking cloud of dust. It entered and inflamed their eyes, it clogged their mouths and made them cough violently, and it terrified the lord's horse so that it reared and whinnyed in fright. The lord was greatly outraged and his retainers indignantly dragged the unfortunate fool from the tree and thrust him to his knees. The man grovelled and whined despicably and beat his forehead on the ground and wept bitterly.

'I have been evil and wicked,' he cried abjectly. 'I have killed my neighbour's dog in a fit of rage and smashed to pieces his beautiful mortar. There has been nothing but envy and covetousness in my heart, and because of them I have greatly wronged my good neighbour. Now I have insulted my lord. Forgive me! Forgive me! If only you will, from this moment I shall mend my ways and my evil thoughts. Only I pray give me another chance.'

The lord was still very angry. He severely reprimanded the ill-tempered man, but at last he forgave him on the condition that, if he did not change his ways from this very day, he would be severely punished.

As the weeks and months passed the old couple grew more serene and happy, and their good fortune for ever grew. Their neighbour and his wife slowly changed their character and their ways. Their envy gave place to kindness; their ill-temper to gentleness; and their unneighbourliness to a warm lasting friendship with the old couple. On every festival and anniversary, all four went to the temple and to the grave of Shiro and offered prayers and mochi-cakes for the everlasting peace of his spirit, and the remainder of their days were spent in generous goodwill to each other and to the people of the village.

The Young Urashima

JUST as the twilight was falling on the village of Mizunoe in the province of Tango, a fisherboy of the village called Urashima was drawing his boat up the pebbled shore after a long day at sea. Young though he was, his skill with sail, hook and line, equalled that of the best fishermen of the village; and on days when it seemed that the sea was empty of fish and his elders lamented over the poor and unsuccessful season, Urashima never failed to return without something to show for his day's labour.

One day, having secured his boat high up on the strand, Urashima started back with his catch to return home. Where the pebbles gave way to a golden stretch of sand, his attention was attracted by a noisy gesticulating circle of children, who seemed to be mercilessly flailing something in their midst. On approaching closer, Urashima found it to be a large turtle.

'It's my turn to beat the drum,' cried one, and down came a stick on the turtle's back.

'It's my turn,' cried another, and a stock of sea-wrack whistled through the air.

'Now, all together,' they cried, and sticks and seaweed stocks rained down one after the other.

With its dazed head withdrawn under its solid shell for safety, the turtle, too slow and ponderous to escape his young tormentors, remained motionless, suffering the barbed pains that every blow darted through his shell to all parts of his body.

'What are you doing?' cried Urashima, moved to anger at their cruelty and the pitiful plight of the helpless creature. 'Stop this at once! Do you think that you are good children to beat this helpless turtle?'

The children paid no heed to him and renewing their blows on the turtle's back said:

'It is no business of yours, Urashima. It is our turtle. We caught it and we can do what we like with it.'

'You have no right to beat it, it feels pain just as you do. Look, if I give you money, will you let me have the turtle?' said Urashima.

'Oh! yes,' they all cried at once. 'If you give us money, it is yours and you can take it as quickly as you like.'

Urashima handed over the money, and the children, with derisive shouts and laughing loudly at his foolishness, ran off to the village. Urashima turned to the turtle and patting its shell said:

'Well, if they had beaten you much longer, your life would have been in great danger. What brought you here, you gentle creature? From now on, please do not be so careless in the way you wander from your native sea.'

Urashima took the turtle in his arms and walked with it to the seashore. Wading out knee-deep, he released it in the clear blue water and watched as it swam and plunged with delight in the waves round his feet. With a look of gratitude at its benefactor, the turtle turned seawards and swam out of sight.

It was three or four days later. The morning was warm
and windless. There was no sound to be heard but the
cry of a seagull gliding overhead. Urashima sat in his
boat far out from land, his thoughts as listless as the line
dangling on the waveless surface of the sea. All at once,
he was awakened from his reverie by a voice as sweet as
a temple bell calling:

'Urashima San! Urashima San!'

'Oya! Surely that's my name. It sounded as if someone was calling me. But how can it be? I am alone and
out of sight of land. Indeed, I must be dreaming,'
thought Urashima to himself, and turned his eyes to his
float again.

'Urashima San! Urashima San!' came the voice
again. There could be no mistake. It was his name that
was being called. He turned quickly round and there,
close to his boat, its head above the glassy sea, was his
friend the turtle.

'Was it you calling me just now, Turtle Chan?' asked
Urashima in great surprise.

'It was indeed, my dear friend,' answered the turtle.
'I have come to thank you for your great kindness to me
the other day and to show you my gratitude and to
salute my preserver—for so I shall always regard you—
with all my reverence.'

'It was truly a small deed,' said Urashima. 'Too small
to remember and not worthy of your warm gratitude.
But please do not wander too far from your home. It
is dangerous and there are always those who will do you
injury.'

'Aye! You are indeed wise, Urashima San,' replied
the turtle pensively. 'The frog should keep to his pond
and the cicada to his treetop. I was foolish. But now I
have learned my lesson, and from now on I shall always
keep to my ocean. Urashima San, I have something to

ask you. Have you ever heard of the palace of the
Dragon Princess?'

'I have heard of such a place,' answered Urashima.
'But nothing of the Dragon Princess, nor have I ever seen
her palace.'

'Then I have a great treat in store for you, Urashima
San,' said the turtle. 'I wish to invite you to the palace of
the Dragon Princess whither I will escort you on my
broad back.'

'Do you really mean that you know the Dragon Princess herself?' asked Urashima in great surprise.

'Not only have I the honour to know Her Highness
well,' replied the turtle with turtle dignity, 'but I am
one of her chief retainers. I told my lady how you saved
my life, and she is anxious to meet you to thank you in
person. So, Urashima San, will you not come?'

Urashima, still not recovered from his surprise at this
strange encounter, replied with some hesitation:

'A princess of such fame, I would, of course, be honoured to meet. But where is her palace? How can I
get there? And does she really wish to see a lowly
fisherboy like myself?'

'Urashima, my lady, as I told you, is very desirous to
show her deep gratitude. She asked me to seek you out
and deliver her invitation. As for getting there, do
not worry. I shall take you on my back and swim with
you along the lanes of the sea which lead to my lady's
palace. It is a wonderful journey and we shall be there
in no time. Come, Urashima!'

So saying, the turtle swam towards him, and steadied
himself against the gunwale of Urashima's boat.

The turtle's words dispelled Urashima's doubts, and
climbing over the side of his boat, he mounted its
back. Immediately the turtle started swimming at a
great pace over the waveless sea towards a pine-covered

rock which seemed to have appeared at that moment, for never had Urashima seen it before. As the rock began to loom high above them, the turtle suddenly dived and moved with graceful and majestic speed into the green depths of the sea. As they plunged ever deeper, they were joined by the fair and lordly fish of the ocean.

First came a detachment of sword-fish which swam far ahead, tunnelling a path through the depths for the turtle and his honoured passenger. Streamers of silken foam flowed in their wake and were borne in long lines by the looped tails of myriads of sea-horses. A school of dolphins followed the sword-fish bearing on their backs fish from distant sea-beds whose phosphorescent scales illuminated the path with multi-coloured lights. A regiment of noble sea bream flanked the sides of the procession, and above and below, in long lines, swam sardines and angel-fish, goldfish and flying fish, globe-fish and tunny fish, cuttle-fish and lamprey, mackerel and herring, and, surmounting all, clouds of transparent jelly-fish.

Down and down the cavalcade swam, until suddenly in a bright flash a castle appeared, lit with thousands of iridescent

foam bubbles; before Urashima's eyes rose a pair of gigantic gates, glittering with brilliant colours which shimmered in the undulations of the sea, and through whose portals all manner of strange fish and creatures swam.

At the gates the turtle stopped. Settling gently on the sea-bed, it slid Urashima off its back.

'Please be so good as to wait here for a few minutes,' said the turtle, and swimming through the gates, it vanished from Urashima's sight, only to return almost immediately to where the boy was sitting, absorbed in contemplation of the wonders around him.

'In the name of my gracious mistress, the Dragon Princess, I welcome you to her august palace,' said the turtle in a most ceremonious voice. 'My mistress waits with impatience to greet her honourable guest. Mount again on my back, Urashima San, and I will conduct you to Her Imperial presence.'

Urashima climbed on to his friend's broad back, and with beating heart he was transported through the magnificent gates. Once inside, Urashima found himself in a paradise where all the rainbows of the world seemed to begin and end. Before him wavered mistily the outlines of a palace of magnificent splendour whose delicate tracery of towers, turrets, and pagodas spiralled upwards to the far world's surface. On approaching nearer, Urashima saw that what he had taken for a profusion of golden blooms and blossoms were rows of beautiful maidens, attired in rich brocade dresses, and each with a page, no less beautiful than his mistress, to attend her and to sway the ornamented fan above her head. On looking closer, he saw that each maiden wore bright bands of sea-grass and anemones among her high-piled tresses; and nestling in the waves of the hair at the front was a young sea bream, while twined among the topknots of the pages small squibs and octopuses wriggled their tenuous limbs.

As Urashima stood in rapt wonder, the lines of attendants parted like a wave in the centre to reveal a young woman of godly beauty who moved slowly towards him. It was the far-famed and legendary Princess Oto, the Dragon Princess. Urashima fell to his knees and bowed deeply.

'You are welcome to my humble dwelling,' said the princess, 'more perhaps than anyone in the kingdom of my sea and beyond it. For you saved the life of my dear and esteemed retainer, and I owe you an eternal debt of gratitude. I and my people will rejoice, if you will honour us with your company for as long as you may stay.'

Urashima bowed deeply again. Then rising to his feet, he walked with the princess along the great corridors of the palace followed by the maidens and attendants. The floors were covered in agate, from which rose coral columns to support the curved roofs and ceiling of coral tracery. From the alcoves along the corridors came strains of music which followed them as they passed; and the waters everywhere were scented with the richest of perfumes. In the room which they finally entered there was a low red table covered with a cloth of richest damask, and two carved chairs of the same vivid red wood.

The princess proceeded before Urashima and sat with queenly grace in one of them. She motioned Urashima to the place beside her.

'You must be hungry, Urashima San, after your long journey; so let us eat,' said the princess, and gestured in command to one of the attendants.

Immediately from between the coral columns came a long line of other attendants, bearing before them on trays of gold lacquer rich dishes from the four corners of the ocean. While they ate, maidens performed dances from the courts of kings of ancient times, and sang melodies of love from tales of old romances to the accompaniment of harp, flute, and drum.

The meal ended, the princess invited Urashima to accompany her on a tour of the palace. They passed through rooms with walls of ivory and blue marble, jade and amber, sandal-wood and cedarwood, and floors of stone from the quarries of distant seas whose colours glowed and fused with the rich hues of the walls. And carved on the ceiling of every room was the magnificent gold and red dragon of Princess Oto's house.

At last they came to a room overlooking a curving red bridge which hung over a stream, deep and crystal clear. Here the princess paused, and standing near one of the sliding screen windows, said:

'Pray, rest awhile, and I will show you the scenery of the four seasons in the space of a few moments. First, we shall look through the eastern window.'

The princess slid back the delicate screen, and before him Urashima saw a landscape in all the freshness and greenness of spring. A haze hung over a grove of cherry trees whose buds were already coming into bloom. Willows wept over the waters of the stream, and from every branch came the song of the little bush-warbler. Urashima's only wish was to be left here for ever, but the princess drew him to the southern window, and opening it, bade him look.

Suddenly summer in all its warmth broke forth. Fragrant hosts of white gardenia blossoms encircled a pond, whose surface was covered with floating lilies of every shade, their petals hung with quick-silver beads of dew. Jewel-plumaged water-fowl darted over the surface, bringing the dew-drops down in a tingling cascade as their wings brushed the petals. Cicadas filled the air with their songs, and frogs croaked in lazy contentment. But now the princess again drew Urashima away and bade him look through the western window.

Before him, the landscape unfolded into the distance and was afire with the autumn red of the maple tree. On mountain and hill, by lake and river, in valley and plain, the earth was laid with a carpet of flames. A mackerel sky hung over the mountain peaks, and the waters of the river and the lake glowed red in the autumn air. And, strangely, the odour of chrysanthemums pervaded the scene, but no blooms could be seen. Urashima, lost in wonderment, was brought to himself again by the voice of the princess who asked him to come to the northern window. As she drew the screens apart, Urashima became spellbound.

It was winter, and everywhere the world lay hushed under a mantle of snow. Twilight floated over the frozen pond, where, single-limbed, the cranes stood in sleep. Bullrushes and reeds crackled in a wind that suddenly rose and died. Tree, bush, and shrub were bloomed with snow, and spikes of frost hung from branch and leaf. Antlered deer idled under the cold, erect pines, and bears, dark brown against the winter white, were having their evening meal of bark.

Urashima's delight knew no bounds. All thought of returning home left his heart; his only wish was to remain for ever with Princess Oto in her charmed, magic fairyland. Month after month Urashima lived in this enchantment. Each day brought forth some new marvel to delight him, and each night some new miracle to entertain him. How long he stayed he did not know, nor did he care. But quite suddenly thoughts of his parents began to trouble him. He grew quiet and sad, quite

different from the gay and happy youth he had been.
One day the princess asked gently:

'Urashima Sama, why are you so sad and distant
from me? What has happened to you? Can we no longer
give you pleasure?' But Urashima turned his face away
and would not answer.

The princess, much troubled, sought every day to
produce newer delights and greater pleasures for him, but
all to no avail. No matter how delicate the rare foods,
how divinely sweet the voices of the singers, how graceful
the dancers or bewitching the princess, Urashima refused to be comforted. At last, after the princess had
again begged him to tell her his troubles, Urashima held
his sleeve before his eyes and answered:

'For some time past I have been troubled by dreams of
my parents. I fear for their welfare and I would dearly
love to see them.'

At this the princess wept bitterly and Urashima, himself deeply moved, took her hand in his and said:

'Do not weep! I wish only to reassure myself that all
is well with them. Only allow me to visit them for a
brief while, and I shall return to live happily with you
for ever.'

The princess was overcome with grief; but seeing
Urashima's unhappiness, she knew that it would become
greater if she detained him.

'Urashima Sama, grief-stricken though I am, I understand. Please go to them. But before you start, I have
something for you which you must take with you.'

With these words the princess disappeared into an
inner room, but returned almost at once carrying a
beautiful gold-leafed lacquer box tied with tasselled red
cords. Bowing deeply, she placed the box before Urashima, who took it in both hands and held it to his head
as a token of his acceptance.

'Urashima Sama, this is a special, a very special box,' said Princess Oto. 'It contains a treasure of untold value, but one which is best left undisturbed by prying eyes. We call it a "*sayonara* gift", and here I give it to you and wish with all my heart for your speedy return. Go at once, Urashima, and we shall wait longingly for you to come back.'

The princess bowed again and withdrew a few steps, hiding her eyes behind her long sleeves and unable to restrain her tears.

Urashima, too, was sad at the thought of leaving his beautiful princess, but knowing that it would be unmanly to show his feelings before her, he answered bravely:

'I will go, my princess. Your farewell gift I will jealously guard until my return. My eyes will never look upon its contents. My only longing will be to see your dear face again.'

He gazed on the box with such intensity that the princess knew he was fighting to restrain his own tears.

'One day you will return to me, Urashima Sama,' said the princess, 'and I shall always be waiting for that day. Take good care of my gift and it will bring you back safely to your home under the sea. But remember, Urashima Sama, do not, for your sake and my sake, ever open the box. It breaks my heart to think of what will happen should you disobey my warning. Let these be my last words to you, Urashima Sama. Farewell!'

The princess was too distressed to see him to the gates. She remained where she was and watched him walk slowly out of the room to where his friend the turtle was patiently waiting. With the box held firmly under his arm, Urashima mounted the turtle's back and they swam slowly through the green depths. Urashima gazed with longing and sadness upon the place which held everything most dear to him, until it grew dimmer and finally disappeared.

Soon the greenness gave place to deep blue, and all at once they rose to the surface on the crest of a large wave which carried them forward at a great speed. In silence they sped along, until at last they came in sight of a low, sandy beach. Urashima's heart suddenly lifted; for here, at last, was home. What a welcome he would have! What wonders he had to recount! The turtle swam to the shallows where Urashima could easily dismount from his back. While he stood in the water, his gift held tightly in his arms, the turtle gently backed away, saying as he went:

'Urashima Sama! *Sayonara! Sayonara!* Please take good care of yourself and I shall be waiting patiently to take you back to your home under the sea. *Sayonara*, Urashima Sama!'

The turtle turned his great body round, and without a further word or look, swam swiftly away.

Urashima watched until his dear friend disappeared into the distance. His heart was heavy and a deep melancholy came over him. He turned to look at his familiar homeland with a subdued spirit. To his great surprise everything had changed and there was not a landmark he could recognize. In bewilderment he walked up from the beach and into the village street. It hardly seemed the same. The temple still stood on the hill, but old familiar houses had been torn down and new ones had been built. The wayside shrine was still at the entrance of the village, but a new road had been cut and a new wooden bridge spanned the river. Everywhere strange faces regarded him with curiosity, and not an old friend could he see. He hastened along the street towards his parents' home, but stood bewildered when he saw it. It was overgrown with weeds; uncut grass grew in tufts where the bamboo gate once stood; the thatched roof was in decay and the walls were crumbling in ruins.

'What has happened?' muttered Urashima as he looked around him. 'Is this my parents' home? Is this my native village? How could this desolation come about in such a short time? Where are my parents?'

At this moment an old woman, whose back was bent parallel to the road, came limping towards him, and Urashima stopped before her.

'Grandma! Where is Urashima's home? Where have his people gone? What has become of them? Please tell me!' he pleaded.

The old woman screwed her head round to peer at Urashima. She looked at him for a long time, and then said:

'Eh! Urashima, you say? Never heard of such a name. I have lived here for over eighty years and I've never come across anyone of that name.'

Urashima became very agitated and said loudly:

'But this is where they used to live. And no one knows better than I. Surely you must have heard of them.'

The old woman settled her old body down by the road-side and remained silent for some time. Many things seemed to be struggling in her wizened head. At last she nodded and muttered to Urashima:

'Urashima! Urashima! That was the name I heard as a child, I believe. Was he not the boy who went over the sea on a big turtle's back and never returned? Wasn't that the boy of the legend? It is said that he was taken to the Dragon Palace and there he remained a prisoner. But I don't know. It all happened so long ago. I heard the story as a child, as I said, and the story goes that it all happened over three hundred years ago.'

Urashima could hardly contain his astonishment and his sorrow as he realized what had happened.

'Three hundred years ago! Three hundred years ago!' Urashima muttered to himself. 'Three years only I thought I stayed, but now it seems that for every year I dreamed, a hundred years passed by. That explains all: my parents dead; our house in ruins; the village not recognizable. Oh! what have I done!' And he wept bitterly in his sorrow.

After a time, his thoughts turned to his princess and to his new home under the sea. There was his only hope. There was his only happiness. He ran frantically back to the beach and scanned the sea for a sign of the turtle. But nowhere was it to be seen.

'Turtle San! Turtle San! Where are you? I want to return at once. Come to me!' he cried. But only the sighing of the retreating sea through the shallows answered him. He sat down in despair and put the lacquered box, which all this time he had been clutching under his arm, beside him. At the sight of it, he suddenly cried in joy:

'Surely her gift will help me? Surely there are instructions inside the box which will tell me how to return to my dear princess.'

Forgetting the princess's warning, he eagerly untied the tassels and with trembling hands lifted the lid. A cloud of purple mist rose from the box and enveloped him in its folds. As it cleared away, Urashima, to his horror, felt a terrible change come over him. His fresh young face fell into countless lines and wrinkles; his bright eyes became dimmed and bleary; his hair turned snow-white and thin. Cramp overcame his fingers and pains his legs, now thin and shot with thick veins. He tried to rise, but countless years racked his whole body, and he found himself staring into the sand, for his back was bent at right angles like the old woman's and he could not straighten it.

'Oh! what have I done? What have I done? I forgot your words, my dear princess, and rashly opened the box. Now I know why you warned me. You sealed my youth into this box and I alone have lost it. Now all is gone, all is gone,' he cried. Tears streamed down his cheeks. Through his almost sightless eyes he gazed at the sea, but there was nothing to be seen, and only the sea itself cried with him in his sorrow.

The Vanishing Rice-straw Coat

THIS droll story happened in a village in Japan in the days beyond the most remote times that man can remember; so far back, indeed, that no one, not even the storyteller, who passed it down to the ears of his children and their children's children, knows the name of the village or in what province it was located. It was an odd village and the people in it were odder still. To tell the truth, they were a curious lot altogether, and apart from being one and all blessed with extremely weak wits, there was not one that had the shape and size proper and becoming to a man or woman. Some had heads as bald and elongated as a pigeon's egg, others big and round as a watermelon, while yet others were so lopsided that it looked as if they had potatoes on the side of their shoulders instead of heads.

In the village there was one who was a real good-for-nothing scallywag of a rascal. He hadn't even a name, so we will call him Otoko San. He was of a very mischievous temperament and loved nothing so much as plaguing the life out of his neighbours. That was all

right, because they gave him as good as they got; but
when one day Otoko San met a Tengu and decided to
play a trick on him, there was bound to be trouble.

Now a Tengu is a fantastic creature indeed. Its name
means a long-nosed goblin, and in fact it has a nose of
extraordinary length, and on its back it sports a roguish
pair of feathered wings. It dresses in the most bizarre
clothes imaginable, and on top of its head is perched a
small black hat tied with strings under its chin. Added
to that, it is possessed of magical powers. So that only a
fellow without wit enough to keep away from a Tengu
would try to play tricks on it and hope to get away with
it.

Otoko San was one such nitwit, and on the particular
day of the story he was idly fashioning a long pipe out of
a piece of bamboo. First he thought of using it to blow
stones through, then he thought it might make a fine
telescope; and it was while he had got it up to one eye,
that out of the corner of the other one he saw a little
Tengu goblin flying towards him.

'Aha!' said Otoko San to himself. 'Here's where I
have some fun! I'll just see if I can't diddle the little
creature into giving me that fine rice-straw coat he is
wearing.'

Without more ado he began to peer intently at the sky
through his bamboo tube, all the while giving exclamations of surprise. This was too much for the Tengu, who,
like all his brethren, was highly curious. He hopped
eagerly round and round Otoko San and squeakily
begged for a peep through the tube.

'What! Lend you my fine new telescope, which has
been specially made for me? Nothing of the sort! Oh!
what a wonderful sight the moon is, I can see every valley
and plain on it. Don't you wish you could see such a
sight, Tengu San?'

This of course only increased the Tengu's eagerness to look through the tube, and he began to offer first his high wooden shoes, then his shiny black hat, and when that was no use, finally his woven rice-straw coat. This was just what Otoko San was after, and in a minute the deal was made. Off went Otoko San as fast as he could, leaving the poor, deluded little Tengu to determine which was weakest—his eyesight or his credulity.

The moment he was out of sight Otoko San put the coat on, and *piff!* there wasn't a sign of him or the coat. Delighted with himself he danced along until he came to the village street. There he had a glorious time, dodging in and out among the people's legs, knocking their food stalls over, tweaking their noses and generally frightening them out of what wits they had. The stall holders and shopkeepers hid behind their screens and marvelled at these antics that continued in a street that was now apparently deserted. At that moment a proud-looking retainer came stalking down it. He hadn't got very far when, suddenly, a violent tweak of his left ear nearly threw him off his balance, and as he turned round furiously to catch his attacker, he found himself spun round in an undignified circle by his right ear. Sitting down with a thud he glared angrily in every direction at the empty air about him, and frightened as the people were, they burst into roars of laughter at seeing such a pompous retainer behaving more foolishly than the greatest fool of the village.

On Otoko San went until, just in front of him, a sober looking workman came out of a shop. He had just bought some fine new socks with snowy white soles, and as he paused outside, he drew them from their paper wrapping to admire them. What was his amazement to see them suddenly fly out of his hand and go crazily dancing along in the empty air. A young village girl, out in her best summer kimono, stopped to stare, and even as she did so, her parasol was whisked from her hand and went spiralling along with the socks in a gay duet, till suddenly both parasol and socks fell *plop* into a stream at the side of the road.

Now Otoko San came to a fishmonger's, where the women of the village were choosing fish for the evening meal. A fine catch of fresh bream had just been brought in and everyone was admiring their size and plumpness. Suddenly it seemed that the biggest and plumpest of the bream sprang nimbly into life, for it flew up into the air with a leap, and then careered down the street for all the world like a flying fish. This was too much for the villagers, and one and all they chased down the road after it, only to have it flung down, dirty and bedraggled, at their feet.

Otoko San, feeling the need of a rest, quickly returned to his home. Behind the door he removed his wonderful straw-coat and at once became visible along with it, which nearly scared his old mother to death, for she hadn't seen a soul come in at the door.

While Otoko San was sleeping she took the straw-coat down from where he had hung it to give it a good shaking.

'Ha! What a filthy old thing he has picked up, to be sure! I'll burn it before he wakes up.'

She immediately pushed the coat into the glowing oven fire and very soon it was only a pile of grey ash.

When Otoko San arose, his first thought was for the coat. It was nowhere to be seen. When at last his mother confessed that she had burnt it, he was beside himself with anger and immediately started gathering all the ash carefully into a large basket, thinking that there still might be some magical strength left in it. He took it away into a corner of the garden and there he took off all his clothing and carefully rubbed himself from top to toe with the ash. Even this mischievous fellow felt a little strange as he watched himself disappearing before his own eyes; for miraculous as it may sound, that is what happened, and in a short space of time there was not a hair of his head to be seen.

In great delight at the success of his stratagem he went dancing along among the evening crowds. The *sake* shops were beginning to fill, and the delicious smell of the wine soon attracted him to the door of one of them. Slipping inside he sat on the floor near a large barrel of liquor, and seizing a moment when the assembled company had their cups all newly filled, he crept in front of the barrel tap and started to drink greedily. Hearing a strange sucking noise, the company all looked round in surprise, but there was nothing to be seen. Yet the sound continued and seemed to be accompanied, now and then, by what sounded very like a loud hiccup. Then the landlord saw his little dog apparently licking at the *sake* tap, and hurrying forward to stop him, came to a dead stop himself. There, just at the end of the tap, was what looked for all the world like a wet red mouth. And what was more, it was quite unmistakably drinking the *sake*! Drops of the liquor fell from something that began to look like a chin and it was at these that the little dog was licking. To the gathered company's utter stupefaction, an area of quite palpably human skin began to appear round the lips, and soon a nose and a pair of intently screwed-up eyes became visible. In the pool of *sake* that was flowing over from the drip-bowl on to the floor a pair of hands began to take shape, and then, a little farther back, a round patch of something distended and plump . . .

Rousing himself, the landlord let out a yell, at which the spectral face looked up with a great start and cocked a bleary eye at the gaping crowd. With a cry the face rose in the air and went bolting out of the shop, accompanied by a madly waving pair of hands. The worst had happened! The ash was fine in its dry state, but once it became wet all its power of invisibility was lost, and now Otoko San was in a sorry state indeed. The crowd raced after him yelling all the while:

'Come on you, show yourself properly! Where have you put all that *sake* you've drunk? Thief! Devil! Just wait till we catch you!'

By now Otoko San was covered with perspiration, and as the ash mingled with the sweat, his skin appeared in streaks and patches like a half-finished drawing. Panting and really frightened, he came to the bridge over the river and immediately plunged in. Frantically he set to and washed every vestige of the ash from his body and hair, and at last, before the amazed eyes of his pursuers, he crept miserable and shivering out of the water. One, more thoughtful than the rest, threw a kimono about him, and they all gathered round him in great curiosity.

'Ma! We thought you were a devil, Otoko San!' said the village head. 'What tricks have you been up to, to get yourself into this state?'

Otoko San hung his head in shame as he haltingly told of his deal with the Tengu. Then the crowd jeered and hooted with laughter.

'What! Try to get the better of a Tengu? You're crazy, Otoko San! Just you keep your nose out of a Tengu's business, it isn't long enough!'

They roared and slapped their thighs at his discomfiture. And for all I know the little Tengu is also laughing still.

The Tale of Princess Kaguya

I

The Child in the Bamboo Tree

THIS is a story as old as time itself. It happened in a far off province and is about an aged woodcutter who cut down nothing but bamboo trees. Each day as he went to the forest he would be followed by crowds of children who cried:

'Grandpa! Grandpa! What are you going to do with the fine bamboos that you cut down?'

And the old man would reply:

'The strongest pieces are for the bamboo worker in his shop: but the delicate stems I shall work into beautiful baskets.'

'Grandpa! Grandpa! Will you make baskets for us too?'

'Willingly! Willingly! But what will you give me in return? I have no children at my house. . . . Will you come and live with me?'

'No! no! Your house is poor and old, and we don't

love you.' And the children, bursting into tears, would scatter like a cloud of little spiders in all directions.

The old man only smiled, but the children's words wounded him deeply: and one evening in late autumn, when the innocent words of the noisy children hurt him almost more than he could bear, he returned dejectedly home to his wife and said:

'Wife, why is our house not blessed with children to care for us in our old age?'

'That I know not,' sighed his wife. 'Over and over I've prayed to the Lord Buddha to bless our house with a child, but he has never listened to me. What more can I do?' And she wiped her eyes as she spoke.

That evening, after their simple meal, the old couple knelt before the family shrine and prayed earnestly that they might have a child to care for.

A few days later the old man was busy as usual in the forest. Not even the red of the autumn maples could lift his dejected spirits. He worked mechanically, without enthusiasm or pride. The 'kan-kan' sound of his axe on the hollow stems echoed through the trees and over the hills in the still air. The bamboo he was cutting was young and slender and deep green. One more cut and he would be through. He brought the axe down, but no sooner had it cut through than the trunk burst into a shower of light from an inner radiance, which lit up the whole grove around him. The old man fell back in fright and astonishment.

'Ma! What miracle is this?' he cried.

As the bamboo toppled over, he became aware of the sound of singing, at first very faint, but becoming stronger and clearer every moment. He looked round, but there was no one to be seen. Then he realized that the voice was coming from the heart of the bamboo stump. Trembling, he gently cut the bark open. There, inside, rested a tiny figure. As he looked closer he saw that it was a maiden with the sweetest face he had ever seen and dressed in the rich robes of a princess. She it was who was singing so enchantingly; but when she saw the old man, she stopped, and held out her little hands to him with a warm smile. The old man thought he had never seen anyone so lovely. Her face was white and fair as the surf of the sea; her hair, long and black, hung over her shoulders; and the eyes with which she regarded the old man shone like stars. From her body came the soft perfume of a myriad flowers, and the sound of her voice was like a waterfall.

The old man lifted her gently into his hands.

'No one but Lord Buddha could have sent such a gracious child,' said the old man aloud, and he knelt and said a heartfelt prayer to Him who had at last so kindly answered their life-long wish.

He tucked the tiny creature into his bosom and with care cut the bamboo stem, which still glowed with a mysterious light. He set off at once for home to bring his wife the wonderful news. As he approached the hut he saw her at the door and she exclaimed in surprise at his early return; but the old man took no notice and shouted:

'A miracle! A miracle! Quick, wife, quick! Fetch the newest of my new baskets. Quickly, I tell you!'

'Eh! Eh! What's this about a miracle! What do you want a basket for?' demanded the old wife, thinking her husband was surely out of his wits.

'Never you mind. Don't ask questions. Just get a basket immediately, and then I'll show you a great wonder,' said her husband impatiently.

The old woman hurried into the hut and soon emerged with a handsome basket. From the breast of his kimono, her husband carefully took the bamboo child—for so he thought of it—and placed it tenderly in the basket.

Followed by his astounded wife, he carried the cradle into the hut, and as soon as he entered, the room became irradiated by a light streaming from the basket; and again there rose from the child's body the lovely flower-like smell.

The old man told his wife of the miraculous way in which he had found the little girl and pulled his nose as long as a red-nosed devil in explaining how he first smelt the flower perfume. The eyes of the old woman grew as round as rice bowls as she listened, and both the old couple were overwhelmed with thankfulness and happiness. At last they had the child they had so long wished for.

Some time later they realized that they had no name for her. What a problem! And how could they think of one suitable for such a radiant creature? Long they thought and pondered, but it was no use. Finally the old woman said:

'Husband, we are simple people and it is not likely that we can think of a name for our miraculous daughter. Let us go to the teacher who lives nearby, tell him the whole story, and ask his advice.'

The old man immediately agreed, and tucking the child in the breast of his kimono, he set off with his wife for the teacher's house.

The scholar was exceedingly interested in their story, though his wisdom was such that nothing could ever surprise him. He contemplated the child for a long time, while all sat in silence. At last he suddenly tapped his knees and said:

'The child is obviously well-born. So well, indeed, that she must be a princess. Since she is so radiant and beautiful, let her, therefore, be called Princess Kaguya.'

And he wrote the name carefully with his finest brush on a paper scroll.

'How can we thank you and repay you for thinking of such a fine name?' asked the old man.

'I do not need thanks,' said the teacher; 'but I will give you some further advice. Do not talk or gossip of this miraculous happening. Keep the child in your home and do not speak of her outside. By following what I say, you will remain free from anxiety. Goodbye to you both, and remember my words.'

The old couple returned home rejoicing over their new happiness and delighted with the name chosen for the child.

Time went by. The couple looked after Princess Kaguya with all the means at their humble disposal. She grew taller and lovelier with every passing day. Mindful of the teacher's words, they never spoke of her but went their accustomed way; and Princess Kaguya herself seemed content to remain always within the house far from the sight of other human beings. In some strange way the house itself became more beautiful. Always the wonderful radiance followed her: the room was filled with her mysterious glow, and the very walls and roof became permeated with the flower perfume of her body. But outside the hut all remained the same, and the neighbours had no idea of the secret life of the old couple.

Four, or perhaps five, years passed in this way and already Princess Kaguya had grown into a maiden as pure in beauty as the moon upon a green hill. The old man went as usual to the grove each day; since he had found Princess Kaguya, his luck seemed to have increased daily and he never found any scarcity of the finest bamboo trees. One day, cutting away as usual, he heard a sudden tinkling noise from the stem, and before his astonished eyes a shower of golden coins flowed out.

'Ara! Ara! What is this?' exclaimed the old man, and gathering up the money he hurried back to his hut.

From that time forth, when his axe struck the young bamboos, the same thing happened, and soon the old couple became rich and prosperous. They were able to buy handsome kimonos and lay new straw mats on the floor of their hut. The old man had no longer any need to continue his labours in the forest, but was able to devote his time to restful pursuits and to the bringing up of the young princess. Naturally, these changes did not go unnoticed, and the neighbours were not slow to talk among themselves.

'What can have happened to our neighbour—he who was so poor and is now so rich and fine?' Tongues wagged and the rumours flew round like fire sparks in a high wind.

'I've heard it whispered,' said one, 'that a beautiful young maiden is hidden within the house. She is the daughter of a high born person, who, for some reason, does not wish her under his roof and has asked the old bamboo-cutter and his wife to bring her up. Undoubtedly it is for this that he has received a great deal of money.'

'Eh, indeed, there may be some truth in your story; for certainly the old man seems to have some very profitable arrangement,' agreed another.

Such idle gossip and rumours spread rapidly and everywhere the old man went the villagers would greet him with sly hints and insinuations, calling him 'Honourable Sir' and 'Noble Lord'.

The day arrived when the old man decided that it was time Princess Kaguya had ornaments to deck her flowing black hair. Off he went to the shop, where he was soon surrounded by a crowd of people who marvelled at the pieces of gold he was able to produce for the trinkets.

'You have become a most wealthy person of late, have you not, neighbour?' they said sarcastically. 'Do you not feel uneasy in your conscience at possessing so much? And who, tell us, is the beautiful girl that it is said you conceal in your house?'

In this way the villagers expressed their curiosity—and at times their envy too. They made the old man miserable and he grew disturbed in his mind.

II

The Fame of the Princess

Soon it became evident to him that the entire village was aware of the existence of a beautiful girl in his house. So he said to his wife:

'Wife, there is no longer any point in hiding our princess, for everyone knows that she is here. Indeed, now that she has become a young lady, it is time she learned more of the world and of the people round her. We will hold a celebratory party for her and invite all the village. It will stop their wagging tongues and give us peace of mind.'

'That is a splendid idea!' said his wife. 'And as we have so much money to spare, we shall give a grand banquet for everyone.' And immediately she set about planning in her mind all the good dishes she would prepare.

Everybody was invited, and in the days that followed a constant stream of callers, from far and near, came to the house to thank their good neighbours for the gracious invitation to the Honourable Princess Kaguya's celebratory party. Meantime the old woman and a band of willing helpers were busy day and night preparing the feast, and the kitchen was in a constant bustle.

At last the day of the party came, and the old man and his wife welcomed their guests with every ceremony. It was the greatest gathering ever to have been seen in the village, and not an inch of straw matting could be seen for the crowd seated on the floor.

Before the main meal was served, the old man rose and announced with quiet, serious dignity:

'Dear friends, I know that you feel I have been hiding many things from you and that I have been acting in a most unneighbourly manner. If I have, please forgive me. It could not be otherwise. But now I wish to take you wholly into my confidence. I know that it is rumoured in the village that the young girl whom we have brought up as our own child is the daughter of some high personage and that my wife and I are her guardians. This is not so. The truth is far more wonderful.'

With this the old man told them the whole story, from the time he cut the bamboo in the grove and found the little girl inside it, to the miracle of the gold coins. The guests were astounded at his story and all begged to see the lovely princess. The old man rose and slid back a screen at one side of the room to reveal, behind a thin silk curtain, Princess Kaguya, looking so shy and gentle in her young beauty that the guests were struck dumb with wonder.

The old man was greatly relieved to have his secret open to the world, and joyfully exclaimed:

'So! Now you have seen the great treasure of our hearts. Come along! Come along! Pray help yourselves and refresh yourselves without ceremony!'

The guests needed no urging and soon the little house was echoing with the sound of laughter and revelry. The old woman and the other wives were kept busy pouring out the hot spicy *sake*, and as its warmth flowed through their veins, the guests grew merrier and merrier. *Sake* cups were exchanged for long friendship; toasts were drunk to Princess Kaguya and the old couple; and for three days and nights the party went on. Not a stroke of work was done in all that time, and only the sound of singing, laughter, and music echoed through the village.

Long after the party was over, praises of Princess Kaguya were to be heard on every side, and rumours of her beauty and miraculous birth flew around the country-side. Soon the story reached the neighbouring towns, and gallant young men began to make pilgrim-

ages to the little village to see the fabulous princess whose beauty was praised so highly. Day after day groups of curious people gathered in front of the gate and even climbed into the garden to try and catch a glimpse of her. But Princess Kaguya stayed always behind her curtain away from their curious eyes, and showed herself to no one. The old man became more and more angry at the discourteous behaviour of the importunate crowds, and finally driven to distraction, he constructed a high earthen wall around the house.

'Please get a very strong lock and key for the gate!' said his wife anxiously, for she feared that the headstrong young men would try to break into the house.

One day an ardent young man, trying to scale the wall, found a crack which he broke open into a small hole, through which he spied on the princess's window. But though at times he could hear a sweet voice singing, not a sign of the princess could he see. He told his friends, and each day they gathered round the hole with their ears pressed against it, listening for the voice which never failed to enchant them. Then one morning when the old man was going out the young men caught hold· of him and begged him to let them see the princess. The old man shook his head vigorously and said:

'The princess is too young and you cannot see her.'

And he refused positively to let them in, adding that such a young and obviously well-born maiden might have great expectations elsewhere. The young men were greatly disappointed and went away grumbling, saying:

'Well there's certainly no chance for us if that's the case.'

And as time went on they gradually ceased their importuning.

III

The Five Young Men

There were, however, among the youths, five young men of high family and great wealth named Prince Kurumamochi, Prince Ishitsukuri, the Minister Abe-no-Miushi, the Chief Councillor Otomo-no-Miyuki, and the Deputy Chief Councillor Iso-no-Kamimaro. Despite the differences in wealth and rank between them, they were all great friends. They had refused to be discouraged by the old man and had managed with each other's help to scale the wall and enter the little yard. There they stood, despite wind and rain, and swore they would not leave until they had seen the princess.

The old man, seeing their plight and marvelling at
their devotion, was filled with apprehension for them and
begged them to return to their homes. But they would
not listen to him and only said:

'Grandpa, do not say such things. We beg only to
remain here until we have seen Princess Kaguya.'

The old man grew more and more disturbed and at
last he said he would speak to the princess. He went to
her and said:

'My dear child, ever since I found you in the bamboo
grove, you have been as a daughter to us and we have
done our best to give you a happy loving home. With
that in mind please listen to what your father wishes
to say to you.'

'Yes, indeed, dear Father,' answered the princess.
'You have done everything for me, and I shall listen
humbly to your words.'

'My child, I am over seventy and an old man now,
and at any moment the good Buddha may decide it is
time for me to leave this life. Before I go, I dearly long
to see you with a fine husband and a home of your own.
Only then can I be at peace. There are five excellent
young men waiting at the gate. They have waited

devotedly to see you, night and day, through cold and rain. One might make a good husband and I wish you to meet them.'

But Princess Kaguya shrank with terror behind her curtain and cried out:

'No! No! I will not be seen! I will not be seen! Tell them to go away at once, at once!' And she shook her head petulantly.

'Well! Well! It is a great pity. They are all of high birth and of long lineage, and I do not know what excuse I can make to them,' replied the old man with a great sigh.

Seeing his sadness, Princess Kaguya was moved to compassion and said to him with a smile:

'Dear Father, I cannot bear to make you unhappy: it makes me unhappy too. I will show myself, but only to him who will bring to me the object I shall ask of him. Take this and read it to them. On it are written my requests.' And she handed the old man a scroll.

The old man was overjoyed, thinking it an excellent device for choosing one of the young men. Meantime they, still waiting patiently, were whiling away the time by playing on the flute, singing, and making poems in praise of the Princess Kaguya. When the old man came out to greet them they stopped at once and waited impatiently for him to speak.

'What did she say? Will she see us? Has she sent any word to us?' they asked eagerly.

'Princess Kaguya thanks you for your constant attendance at our poor house,' said the old man, 'and if you will do as she directs, she will come in person to greet the one who first accomplishes her wish.'

'Tell us! Tell us! What must we bring her? We are ready to go to the farthest curve of the earth or sky to carry out the princess's wishes.'

The old man produced the long scroll on which Princess Kaguya had written her commands for the five suitors. The first was for the young Prince Kurumamochi. He was to go to the Horai Mountain and bring from there a branch bearing a gleaming white ball which he would find growing on a golden tree.

'A branch bearing a gleaming white ball growing on a golden tree!' repeated Kurumamochi a little bewildered. 'Surely that is a tree no human being has ever set eyes on.'

'As for you, Prince Ishitsukuri, you are requested to find the stone bowl used by the great Lord Buddha himself for drinking when he journeyed through the world,' continued the old man.

'Oh! but such a thing is impossible,' lamented poor Ishitsukuri.

To the Chief Councillor Otomo-no-Miyuki, the old man said: 'Princess Kaguya requests that the Chief Councillor Otomo-no-Miyuki bring her the Ball of Five Jewels which you will find in the throat of the Dragon of the Horai Mountain.'

'As for you, Iso-no-Kamimaro, she desires you to bring her the Cowry Shell which the Horai Mountain Swallow bears within her. Nor must you harm the bird nor the Shell in obtaining it.'

'The princess is certainly exceedingly hard to please,' groaned Iso-no-Kamimaro when he heard his task.

To the Minister Abe-no-Miushi fell the last command. He was to bring back the skin of the Tree Rat which lived in the mountains of China and which was reputed to vanish into thin air at the slightest flash of danger, and whose skin had this further miraculous property, that however glowing the furnace or burning red the fire, it would emerge uncharred and unscathed from their flames.

All the young men sighed despondently and remained silent for some time, each lost in his own dejection.

'How can Princess Kaguya expect us to do such impossible tasks?' grumbled Abe-no-Miushi and fell to silence again.

They turned away from the house, and on the way home they played their flutes and recited poems about the difficult tasks, trying in vain to keep their spirits up.

IV

The Task of Prince Kurumamochi

When he reached his home, Kurumamochi said to himself:

'Since I am quite certain that the Horai Mountain can never be discovered and that there is no such thing as a gold tree bearing a gleaming white ball, why should I not get my retainers to make a ball and branch for me?'

Excited by his idea, he set up such a tattooing on his drum that his retainers came rushing in from all directions.

'I am commanded by the Princess Kaguya to seek and bring to her a branch bearing a gleaming white ball of the Golden Treasure Tree which grows on the Horai Mountain, if I am to win her favour,' said Kurumamochi. 'Which of you will be willing to accompany me?'

This was an adventure exactly suited to the young men of Kurumamochi's house, and they lost no time in preparing for the journey. There were many sighs of foreboding and pleas for caution on this dangerous mission from among the many retainers who came to see Kurumamochi and his train embark on their gaily decorated ship; nevertheless the adventurers left in high spirits.

Many days later they arrived at a quiet shore on a remote sea. The young men were somewhat crestfallen to find that what they had imagined to be a journey of adventure and daring was, in reality, an occasion for turning them into temporary workmen. For after revealing his plan to them, Kurumamochi called his young men to a secluded glade in the shade of the mountain and laid a vow of secrecy on them, making them promise never to reveal what he was about to do. First, they were to build a high fence so that no prying eyes could peep. Next, they must bring to Kurumamochi trunks and saplings of the finest trees they could find, together with the most delicately leaved branches. To his followers he said:

'Princess Kaguya desires a branch from the Golden Tree bearing a white ball. I do not believe such a tree exists. But, as I am determined to win the princess's favour, we shall ourselves construct a branch bearing a white ball. It is for this purpose that I have asked you to come with me. Now, let us begin.'

A year passed, and still they worked. Day and night nothing was heard but the 'kotsu-kotsu' sound of planes and chisels, and the 'ton-ton' of mallets. During this time their supplies were plentiful, but as the second year passed they began to dwindle. The men became exhausted and grew lean with hunger. Yet, until the work was finished, they could not leave the mountain.

The day came when the last shining leaf had been polished, the last delicate lick of gold applied, and the ball shone so brilliantly white that it dazzled the beholder. They could not have gone on much longer, and great was their joy as the high fence was pulled down and they were able to regard the outer world again. They came down from the mountain to the sea where they caught fish and gathered sea-grasses and seaweed. They ate their fill, and it was not long before they regained their health and their bodies filled out.

Kurumamochi was delighted when he looked upon their work and promised gifts and rewards to all his men when they returned home. The retainers were overjoyed and said that the finished branch and ball were indeed miraculous treasures, since they were to bring such good fortune to all.

With rejoicing the young men returned to their ship, bearing the glittering branch before them. The weather was fine with a strong wind, and soon they reached the shores of their homeland, where news of their approach had preceded them and where a great crowd was waiting to welcome them. Everyone exclaimed in wonder at the beautiful branch. They were convinced that it was indeed from the fabulous Golden Tree, and Kurumamochi grew more and more pleased with his bold idea. He ordered the branch to be placed in a gold lacquered box, and bidding his helpers remain and rest themselves after their labours, he called for his horses and his servants and set off for Princess Kaguya's house.

When the old man heard the procession arriving at his gate, he went to see who it was and was astonished to see that Kurumamochi had returned. He had thought that all the young men had either given up their tasks, or had perished in the attempt, and he could hardly believe his ears when he heard Kurumamochi say that he had brought back safely the golden branch from the tree on the Horai Mountain. He conducted the young man inside the gate and into the house, where he asked him to wait while he conveyed the news of his arrival to the princess. Kurumamochi smilingly assented. He was full of confidence and elation, and he conducted himself without reserve or ceremony, reclining on the cushion as though he were already an accepted suitor.

The old man, well accustomed to carrying heavy loads of wood, found it no effort to lift the large box and carry it single-handed into Princess Kaguya's room. Greatly surprised, she demanded to see inside the box at once, for she had thought her requests were quite impossible of fulfilment. But her surprise turned to awe when, as the old man lifted the lid, a brilliant stream of light flowed from the branch; and when he lifted it out of the box, the entire room was lit by its radiance and filled by the sweet sound of bells coming from the gleaming ball.

'A beautiful sight! A truly glorious sight!' cried Princess Kaguya. 'Surely it is from the Golden Tree itself.'

She looked at it attentively, and suddenly her face grew clouded; and the more closely she examined it, the more clouded her face became.

Meanwhile the old woman, hearing her exclamations, came running in. She too began adding her praises and was delighted to think that the princess would now become betrothed to the successful young prince. But when she reminded the princess of her promise, she refused to listen and sat quiet and unhappy.

Meanwhile Kurumamochi had become impatient, and throwing the demands of courtesy and custom aside, he came unannounced into the room. Princess Kaguya hid herself even further behind her veil as Kurumamochi said:

'Princess Kaguya, I have returned. I

have brought back a branch from the Golden Tree on the Horai Mountain. This was your request: and did you not promise yourself to the one who should fulfil your request? I have come to claim your favour. What is your answer?'

He would have approached closer to her, but the old man stopped him and said:

'But how and where in the world did you get this wonderful branch, Kurumamochi Sama?'

Kurumamochi drew himself up proudly and gave a loud 'Ahem' before proceeding:

'After Her Highness conveyed to me my task, I returned to my home. I made preparations for the voyage and soon boarded my ship. I sailed many times in different directions, having no idea where to look for the Horai Mountain. After we had been at sea for several weeks, a great storm blew up, and for what seemed like five hundred days we drifted at the mercy of the waves. But one day the sea became quieter and we put in at a narrow shore at the foot of a high mountain. Here we made a camp and recovered from our perilous voyage. One day, as I was exploring the hillside, a beautiful young woman appeared before me. She was carrying a bucket in which gold was immersed in crystal clear water. I saluted her courteously and said:

' "What is the name of this mountain, Lady?"

'You can judge of my astonishment and delight when she told me that it was called the Horai Mountain. I set off for the summit at once, where I felt sure I should find the Golden Tree. I spent many weary days struggling up the steep slopes, but, after the tenth day, I reached the top and found myself in surroundings of incomparable beauty. The ground was carpeted with flowers; in the sky sunset clouds glowed over the scene; and birds of every colour and song flew overhead. As I

144

walked on, I came to a river in which the water shone like silver. It was spanned by a golden bridge. I looked across it, and there, on the other side, stood a wonderful golden tree, hung with many glistening white spheres. "Ah," I said to myself, "that is certainly the Princess Kaguya's famous tree." I ran over the bridge, and stood for a long time wondering at its heavenly radiance before breaking off the branch you see here. Never shall I forget the music of its bells, as I broke it from the tree; it filled the whole valley and brought with it beautiful dreams. Then I hastened down the mountain and rejoined my men, and we at once set sail for home. For what seemed like four hundred days we sailed on a heavy sea with a strong wind; and only yesterday we landed. Wishing only to see my princess, I paused neither for sleep nor food, nor even to change my travel-stained kimono. It has been a hard test, and I, with my brave men, grew thin and weary. But now I have accomplished it and I have come to claim the hand of Princess Kaguya.'

Kurumamochi delivered this lying tale with the greatest composure, while the old man and the old woman nodded their heads and sighed with admiration and amazement. But when he turned towards the princess, she kept her head bent down and refused to look at him.

Seeing her reluctance, Kurumamochi became more peremptory than ever. Striding to the outer room, he summoned his retainers and bade them in a loud voice to set off for home at once and begin preparations for the wedding ceremony.

'And let no expense be spared,' he cried, 'for such a hard won princess must surely merit the most lavish entertainment.'

He cast an angry look at the princess in the inner

room as he said this, but she only shrank even closer behind her veil.

Suddenly, there was a loud clattering of hooves outside, and up to the gate galloped five or six young men, the foaming mouths of their beasts and their own excited looks betokening their haste. They rode through the open gates into the courtyard without ceremony and, dismounting, rushed up to the inner door.

'Let us in! Let us in!' they shouted. 'We come with a letter for the Princess Kaguya. It is most urgent, and into her hands alone will we deliver it!'

Kurumamochi rushed out when he heard the noise, and seeing the young men, his face turned white with anger and apprehension.

'How dare you come so unceremoniously into this house! Where is your respect? Do you not know that this is the occasion of my betrothal to the Princess Kaguya? Be off with you and trouble us no further!'

So he berated them, but Kaguya's old father checked him.

'No! No!' he said. 'If they have indeed a letter for my daughter, it is only proper that I should receive it for her. I will question these young men. Pray be patient!'

But Kurumamochi thrust him aside and shouted in great rage:

'These fellows show the greatest incivility! They have neither manners nor respect! They came here like a noisy, ill-bred rabble! How can you listen to them!'

'Aha! Just ask us what we know about that lying fellow there, and you'll soon see who is the ill-mannered, uncivil one!' cried the youths, and they pointed their fingers derisively at Kurumamochi. 'We know how he has tried to trick the princess and we demand to be allowed to deliver this letter to her, which will tell her of his evil plan.'

The noise of the quarrelling and shouting had drawn the princess from her room. Standing at the door, she overheard what was said about the letter.

Her voice fell like a soft veil across their anger as she requested the honour of receiving a letter which it seemed some honourable person had deigned to send her.

'That surely is only reasonable,' said the old man, and he took the letter from one of the young men and handed it to the princess.

'When the princess has read your letter,' he continued, turning to the young men, 'she will speak with you.'

'We are content to accept the gracious Princess Kaguya's favour,' answered the young men, bowing. 'She shall decide whether we are right or wrong.'

The princess unfolded the scroll, and as she read it her colour rose and her bosom swelled with indignation; for the letter said:

'We are retainers of the palace of Kurumamochi Sama. We sailed with him, and our travels brought us to a mountain, but it was not the Horai Mountain. We did not find the Golden Tree, for the simple reason that we never looked for it. Instead, he kept us shut away from the world and worked us day and night to produce the false branch and gleaming white ball, which he now presents to you as coming from the Golden Tree in the Horai Mountain. For our labours he promised us rich rewards, but not a coin have we received. We no longer believe his lies. He is a treacherous man and has done a great wrong to your gracious person and to us.'

'This is indeed a sorry story!' cried the princess. 'Such wickedness saddens me. You were quite right to inform me that the golden branch that seemed so fine is only a hollow sham. I shall myself reward you for your kindness.'

She asked the old man to bring money and presents for the retainers. They were overjoyed at the richness of their gifts and returned rejoicing to their homes. Turning to Kurumamochi, Princess Kaguya bowed with dignity and said:

'Since it seems that there is nothing more to say, I beg you to excuse me.'

She retired to her room with composure, leaving only the exquisite flower smell of her body behind her. Kurumamochi would have rushed after her, but the old man stopped him and said:

'You must go from here at once. How dare you practise such a deception on us! Take your precious tree and begone!'

The old man ordered his servants to put the box containing the golden branch outside the gates and to see that they were locked, with the impostor Kurumamochi outside them.

Kurumamochi's rage and mortification at the collapse of his plan knew no bounds. He stamped his feet and kicked the box angrily, and his passion became the more furious as he thought of all the money it had cost him, and to no effect. But, rage as he would, there was no response from within the house, and at last he rode off, angry and disconsolate, to his own Province.

v

The Task of the Minister Abe-no-Miushi

At about the same time that Kurumamochi was busily constructing his false branch, one of the other young men, the Minister Abe-no-Miushi, was racking his brains over his task.

'Somehow or other I've got to lay my hands on this magical ratskin,' he said: 'or at any rate, one very like it.'

He pondered over this last thought and suddenly his face brightened.

'Of course! That's the answer! I'll get one made. Princess Kaguya will never know the difference.'

Highly pleased with himself, he sat down at once to write to Okyo San, a very skilful friend of his. In the letter he enclosed a large sum of money, and having explained what was needed, told his friend that he must use it all if need be in the search for a suitable skin, or in the making of one. His friend came at once in answer to the letter, and as he and Abe-no-Miushi sat together, he said:

'Certainly I have heard reports of this fabulous mountain Rat, but I doubt if such a creature really exists. However, I will first make a wide search. If I am unsuccessful, I will set about making one for you. I am confident I can do it: so please wait patiently until you hear from me again.'

With these words, he went off to prepare for his journey.

The Minister was delighted and said to himself: 'Ha! Once I have the skin in my possession I will place a gift of great price in the hand of this man, for through him Princess Kaguya will become my own prized possession.' And he settled down to wait for the return of his friend with the skin.

One year he waited; two years he waited, and still there was no sign nor word from his friend. When nearly three years had elapsed, Abe-no-Miushi decided to send a letter after Okyo San. But no answer came, and he thought that his friend had surely run off with the money he had given him and had no intention of returning. His anger grew with his impatience, and he was about to set out himself in search of him, when a letter arrived. In it Okyo San said that he had searched in every quarter of the country for a fire-resistant ratskin and that, after experiencing many dangers, he had at last come to a temple high up the sides of the Horai Mountain, and had learned that a priest there had the precious skin hidden away. After months of bargaining, he had at last been able to buy the skin, but he was short of the necessary money by fifty *ryo* and he begged the Minister to send him the amount immediately, so that he could return with the greatest possible speed.

The Minister was delighted with the news and at once dispatched a number of his retainers with the money. After some time they returned with Okyo San bearing a handsome red-coloured box in which was the ratskin. When Abe-no-Miushi took the skin out and unrolled it, he was staggered by its rare splendour. It gleamed with the silvery blue of the sky, and when a breeze, however small, disturbed the gleaming depths of the fur, waves of colour as rich as a peacock's tail rippled across its surface.

'What beauty! What a magnificent treasure! Searching for this must have caused you great hardship and toil, my friend,' he said, turning to Okyo San. 'Accept my heartfelt thanks now. On my return from the house of Princess Kaguya, I will richly reward you.'

Accompanied by his retainers, Abe-no-Miushi set off for Princess Kaguya's house with a great clatter of horses' hooves and jingling of harness. When he arrived there, and the old man heard that the Minister Abe-no-Miushi had returned with the miraculous ratskin, he at once gave orders for his admittance and called the princess and the old wife into the room. Abe-no-Miushi carried in the handsomely ornamented box, and the princess, sitting behind her veil, asked for the skin to be spread before her. This being done, she seemed struck with astonishment and said:

'How really beautiful it is! Is it not, dear Father? What exquisite colours! Yet there is one test before I can be sure this really is the skin of the fabled Tree Rat of Horai Mountain. One of the qualities of this skin is that it cannot be destroyed by fire. So please prepare a fire and place the skin upon it. If it does not burn, I shall know that it is indeed the skin I asked for.'

At these words the Minister Abe-no-Miushi stepped forward and, picking up the skin, said confidently:

'You need have no fears, Your Highness. This is, indeed, the real skin. I will myself place it on the fire.'

The old man bade his servants build a wood fire in the garden. As the flames gathered strength, the old man, his wife, the Minister and his retainers came out into the garden, full of excitement and anticipation. But the princess remained aloof, where she might see, yet not be seen. Abe-no-Miushi stepped forward holding the skin before him. As the flames leaped ever higher, they were bathed in light from the radiance of the skin, which in turn seemed brighter than the fire itself, and the whole garden glowed into great brilliance. For several moments Abe-no-Miushi stood with the skin before him. Then, with sudden decision, he threw it into the heart of the fire. It seemed for a moment to dull the brilliance of the flames, and the retainers raised a great shout:

'What a wonder! It does not burn! It does not burn!'

But even as the words left their lips, a hideous change came over the skin, as, writhing and contracting, it grew black and charred before their eyes, till at last nothing remained of its former beauty but a twisted and blackened shred.

Abe-no-Miushi's face grew pale and he shook with anger.

'What!' he shouted. 'Nothing but a charred rag! And oh! so much money spent on it.'

He stared at the dying
fire and the shrivelled re-
mains of the skin, and rage and indignation slowly
mounted until they almost choked him. But the laughter
of Princess Kaguya was like silver bells, as she passed him.

'Ah, now there is no need for me to go with you, and I
may stay here where I am so happy,' she said, and holding her veil about her, she disappeared into the house.

VI

The Task of the Chief Councillor Otomo-no-Miyuki

But what was happening to Otomo-no-Miyuki all this
time? While pondering on his own task, he heard
rumours of the failures of the other two and laughed to
himself at their stupidity.

'What!' he said. 'Do they really imagine they can get
away with such tricks? Naturally the workmen gave the
game away when Kurumamochi failed to pay them as he
promised. And for the Minister Abe-no-Miushi to be
taken in so easily over the ratskin! Truly, I shall not be
guilty of such knavish folly.'

Although Otomo-no-Miyuki was of very good family,
he was comparatively poor and was forced to live in a
careful and modest way. This now caused him some
concern, for he thought that should he be successful in

winning the princess, she would hardly appreciate the small size and number of his family dwellings and the fewness of his servants and retainers. However, he put this worry to the back of his mind, for if he was to have the princess at all, then he must first organize the search for the Ball of Five Jewels, and to this end he summoned his personal retainers and said:

'In the distant and dangerous Horai Mountain, there roams a giant Dragon who carries in his throat a Ball set with five jewels. I wish to obtain possession of this rare Ball and I want you, my followers, to prepare your-selves for this task and set out at once. Use any means in your power to bring it back to me, and I shall richly reward you. As a proof of my intention, I will give you each a handsome reward at once, and when you return with the Ball, it shall be doubled.'

He gave them a purse heavy with gold. His retainers accepted it with many protestations of unworthiness and gratitude; but once alone, they began to whisper among themselves. Said one:

'This is a dangerous and difficult task our master has put on us. I have heard many tales of the Horai Moun-tain Dragon. It is said that his magic power is so strong that no mortal has ever been able to approach him. How can we hope to fare any better? And if we return empty-handed we cannot hope to receive the other half of the reward. Let us profit by the wisdom of the ancients and make sure of keeping the thing we have, and not risk losing our lives for something more we may never attain.'

As none of them were any more anxious than he to face the dreadful Dragon, the others readily agreed. They divided the money equally among themselves and each went his separate way as speedily as possible.

Otomo-no-Miyuki meanwhile set about enlarging his

house for his anticipated bride and awaited impatiently the return of his retainers. Months passed and still there was no sign of them. At last Otomo-no-Miyuki was forced to accept the fact that they had gone for good. Refusing to wait any longer, he decided to go off himself in search of the Dragon, and to this end he speedily set about building a boat, large enough for himself and a small band of sailors. When it was ready and the sailors hired, he called them together and told them the purpose of his journey. The sailors were at first reluctant to embark on such a dangerous mission, but Otomo-no-Miyuki urged that they had nothing to fear in the service of one who was descended from a noble line of warriors. His argument prevailed and Otomo-no-Miyuki and his crew set out.

They sailed for some months under a fair sky and a soft wind that bellied their sails and drove them along with gentle speed. But as they steered south of the far coast of Kyushu, the sea began to grow hourly rougher; the wind rose to a gale; the waves reared mountain-high; the sea roared and churned in a thousand foaming whirl-pools; and the little boat tossed and turned like a wooden rice bucket. The sailors, who while the good weather had lasted had been in good spirits, were now overcome with fear. But Otomo-no-Miyuki shouted encouragement to them.

'I fear neither the anger of the sea nor the Great Dragon himself. I have the strength and courage of my great line of ancestors. I bear their strong heart which they carried to victory in all their battles, and I carry it now in this battle with the elements.'

But despite his proud assertions, he was forced to cling for dear life to the sides of the boat as it rocked and reeled, and his pride in his ancestors in no way alleviated the pangs of sea-sickness which now began to assail him. It took all his endurance and self restraint to refrain from sliding to the deck in a miserable huddle alongside the sailors, who had no noble blood to live up to.

The storm pressed them violently. The boat was helpless and entirely at the mercy of the sea, until finally the waves swept them against a sandy shore. There they lay for several days and the storm abated. No one aboard escaped injury and sickness. The deck was strewn with bodies more dead than alive, and their groans and cries of pain could be heard above the easing storm. At last, one fine morning, Otomo-no-Miyuki pulled himself to his feet and looking out over the side of the boat murmured:

'We are at least on land and safe. That is something.'

His eye followed the pine-dotted beach to a distant mountain that rose naked from the plain.

'Indeed,' he thought, 'it may be the land of the Horai Mountain itself.'

Suddenly at his words it seemed a high breeze sprang up and swirled the tops of the trees on the shore and twisted the tattered sail of their craft in gusty breaths.

'Perhaps that is the breath of the Horai Mountain Dragon himself,' said Otomo-no-Miyuki, and in his conceit and pride he raised his head as high as he could stretch it and glared at the sky. Immediately, a dull shadow fell over the Mountain, and an unearthly sound burst through the clouds. With a great roar, thunder clashed forth and lightning split the sky as the little boat was again seized in the fierce grasp of the waves and carried seawards. Round and round it spun like a top, and the sailors, too sick and weak to hold on to anything, were dashed to and fro across the deck. Otomo-no-Miyuki lost all his bravado, for he had never seen such a sea nor experienced such a storm, and he was sure that all on board were lost for good.

'This is a punishment from the Dragon! Pray to him! Beg him to forgive us all!' yelled the sailors weakly. With all his fine pride gone Otomo-no-Miyuki fell abjectly to his knees and raised his head in supplication.

'Lord Dragon! Lord Dragon! Forgive me, I beg you. Yes, I planned to steal from you the Five-Jewelled Ball. It was a base and evil plan. Only abate this storm and let us go safely on our way, and I vow never to think again of so much as touching even one of your honourable whiskers!'

As well as he could in the storm-tossed boat, Otomo-no-Miyuki bent his head to the deck in token of abject repentance.

The storm began to subside as suddenly as it had arisen, and soon they were sailing calmly on a gentle swell under a bright, serene sky.

'Ha! Lord Dragon has heard my prayers,' murmured Otomo-no-Miyuki joyfully.

After many weeks they at last saw land ahead. Coming inshore they found a small harbour, and here they anchored while they prepared to find out where they were, for they had lost all sense of direction during the great storm.

Otomo-no-Miyuki lost no time in offering thanks to the Lord Dragon for directing them to land; and for the first time in many months his mind was free from anxiety. He set about attending to his men, bringing them food and water from the meagre store that was left, binding their injuries, and giving them encouragement with the news that they would sail away from the Dragon's mountain as quickly as possible. The news gave them more strength than the food and water, and in a short time they had repaired the boat and sails sufficiently for them to be able to sail away.

'Nevertheless,' he thought to himself, 'this is not a very

156

happy state of affairs. We are safe from the wrath of the Dragon; we are on dry land. But what land? Where lies our native country? And even if we ever return alive, the Princess Kaguya can never be mine.'

These thoughts soon dispelled the peace from his mind and filled it with despondency.

But the boatmen were cheerful. They were busily exploring the locality, roaming through the pines on the shore, and scrambling over the rocks to discover some known landmarks. Suddenly one who had scaled a high promontory threw his hands in the air and shouted:

'It is Akashi! It is Akashi! There lies the beach of golden sand and the island rock with two pines! We are home! We are saved!'

The other men ran to his side and looked at the distant golden beach and the twin pines on the rock rising above the horizon, and they wept.

'It is indeed our beloved Japan,' they murmured in joy. 'We are home at last. We are saved!'

On hearing the shouts, Otomo-no-Miyuki raised himself and looking around him in wonder he suddenly recognized the place: the green pines fringing the shore, the soft blue of the misty mountain in the distance, and the shining white sands of his own native land.

'Oya! Oya!' he wept unrestrainedly. 'We are saved! We are saved!'

By and by his eyes cleared and his emotion subsided. The sailors, who now stood round him, chuckled among themselves:

'Our fine warrior lord has two eyes as red as the red plums of spring,' they taunted; 'perhaps he has brought two red balls back in place of the Jewelled Ball of the Dragon King!'

They lost no time in preparing to return to their own

city; quickly they fashioned a palanquin for Otomo-no-
Miyuki. When it was ready, he sat inside it and the
sailors hoisted it on poles on to their shoulders and set
off for home.

Meanwhile, rumours of their doings had reached the
ears of Otomo-no-Miyuki's faithless retainers and they,
hearing of his ill-success, boldly assembled to greet him.

'We also, my lord, tried desperately to obtain the
Jewelled Ball from the throat of the Dragon King,' they
said, 'but we fared no better than your honourable self.'

Believing that they spoke in all honesty, the Chief
Councillor Otomo-no-Miyuki had no heart to be angry
with them. Only when he thought of the severity of the
task that Princess Kaguya had set him and his sufferings
in trying to accomplish it, and then, above all, of the
fine new house that he had built almost single-handed,
especially for her, he was filled with rage and frustration.

'So! You tried, and I tried! Twice we tried to obtain
the Jewelled Ball for Princess Kaguya. We all suffered
greatly and yet with all our efforts we could not accomplish it. Why? Because the request itself was beyond all
reason, and impossible to carry out.'

And seizing a large axe, he started to chop down the

house he had built with great fervour and did not cease
until it was nothing but a shattered pile of firewood.

VII

The Task of Prince Ishitsukuri

When setting Prince Ishitsukuri's task, Princess Kaguya
had written:

'I have always longed to see the bowl used by the first
Lord Buddha. If you can bring this to me, I shall gladly
be yours.'

It was quite certain that the bowl used by the first
Lord Buddha was not to be found in Japan. Prince
Ishitsukuri remembered an old tale that there was a
certain Shaka Temple in India where this priceless relic
was jealously guarded. But India was so distant: it
might be 2,000 miles, it might be 10,000 miles; moreover,
he had heard that the climate was so hot that few
strangers could survive there. And if one became ill in a
country so far from home, what then? After thinking a
long time on these lines, Ishitsukuri sat down and wrote
the following letter to Princess Kaguya:

*'Honourable and Gracious Princess: Today I set out upon
the journey to find the bowl of Buddha you so earnestly desire
and which, if I discover it, will bring to my house a love more
beautiful than the sight of cranes flying homewards at evening.
I will travel far from my native land. Mountains and rivers
will fall behind my back, ten thousand-fold the waves will pass
beneath my feet, the wild geese will have come and gone many
seasons, and the blossoms will have fallen and renewed themselves times without number before my return. I carry no fear
in this dangerous journey, for I am borne onward by a passion
to succeed and by my love for you!'*

So he wrote. In fact, the moment he returned to his own house, he summoned his retainers and said:

'I have an important mission for you to perform. I desire you all without a moment's breath of leave-taking to depart at once, each to a different country, and to bring me back from the oldest and the most famous temples there, the drinking bowl of most antiquity among their treasures.'

Needless to say, his retainers were not at all pleased to be given such a wearisome task, and they grumbled much among themselves as they prepared to set out. However, there was nothing for it but to obey, but they agreed privately to seek the easiest means of satisfying their lord. So, though they went to many temples in their travels, they bothered little about fame or antiquity, but just gathered together as many bowls of every size, colour and texture, as they could carry. In this way the best part of three years passed away. When they returned to their lord's mansion, they had a truly imposing array of bowls for his inspection.

When he saw them, Ishitsukuri raised his hands in delight. Certainly among them he would find the bowl of Buddha, he thought.

'Let me see them one by one,' he cried. And the great procession began, as in single line his retainers came and set their bowls before him. With his hands folded in his sleeves, Ishitsukuri sat cross-legged on his cushion, and each bowl was gently turned and lifted for him to see. As one by one they were rejected, an attendant gestured the retainer away with a negative wave of his fan, and gradually the pile of rejected bowls mounted and the procession of retainers dwindled. Nearly all the bowls had been brought before him and Prince Ishitsukuri's face was growing longer and angrier as he shouted:

'How is it possible that among you all, no one has

brought me the bowl I asked for? These are fine bowls, handsome bowls; but none of them have the age or simplicity of the bowl from which Lord Buddha in his humility doubtless chose to drink.'

And he struck his knee angrily with his fan.

Hearing his words, the last retainer in the procession, who was head of all the others, put down the fine bowl he was carrying and hastened into a side room. Here he had discarded several bowls as too old and dirty for his master's inspection. But now he chose from among them the oldest and dirtiest he could find, and covering it with his capacious sleeves, he again took his place at the end of the line of retainers. When his turn came to present his bowl he stepped forward and, with many deep bows, approached Prince Ishitsukuri.

'My lord,' he said, 'in my travels I came to the famed Horai Mountain in China, and there, after many weary days, I found the renowned temple where the priest lives who alone knew the story of the bowl of Buddha. He told me how it had been brought from the far land of India and had for countless centuries been his temple's greatest treasure. My lord, even a priest may feel the relaxing effect of a little wine, and when we had drunk together he offered, with a great display of secrecy, to show me the treasure. For many days I stayed with him, and when I had completely won his confidence I prevailed on him to sell the bowl to me. Today I have the honour to hand it to my lord.'

With his head bowed to the floor he pushed the bowl gently towards Ishitsukuri.

Ishitsukuri was completely taken in by this and contemplated the bowl in rapt delight, as it stood in all its dingy blackness and dirt before him. Calling for his attendant, he ordered the finest and most rare silk cloth to be brought and had the bowl wrapped in it before setting out with the utmost haste for the house of Princess Kaguya. On the way he gathered flowering branches and garlanded the silk-wrapped bowl as though it were indeed the rarest and most precious of all treasures. With a great number of his retainers he arrived at Princess Kaguya's house.

He was received with all due ceremony, and coming into the room where the princess was waiting, he said to her with an earnest expression:

'I have the honour to bring for your Highness the bowl of Buddha which you so much desired.'

He unwrapped the bowl before her. As he did so a cloud of dust and dirt rose from the bowl and tainted the air. The princess shielded her face with her long sleeve, and lifting her head, rose disdainfully to her feet.

'The true bowl of Buddha is filled with a celestial light: but this bowl is black and dark as ebony. It is not the true bowl: it is false, and I am therefore absolved from any promise to you,' she said scornfully, and glided thankfully away to her own room.

Ishitsukuri flew into a violent rage and stormed and roared so that people fled in all directions.

'The shame! The mortification! A counterfeit bowl!

An imitation! Why have I been deceived? Why, any-way, should she treat me in this way? Oh! What an abominable, detestable girl she is!'

Quite forgetting his own deceitful behaviour in the matter, Ishitsukuri poured abuse and invective on the heads of everyone about him. As he rushed through the gates, they closed firmly behind him, and in the silence left after his thunderous tirade, peace seemed to fall on Princess Kaguya's house and only the sound of twittering birds broke the stillness.

Prince Ishitsukuri turned and gazed at the gates, and suddenly, comparing the joyous expectancy with which he arrived and his bitter mortification on leaving, he was filled again with rage. Seizing the offending bowl he hurled it again and again at the immovable gates, until with one mighty crash it shivered into fragments at his feet. Mounting his horse, he rode furiously away, followed by his subdued retainers and the whispering of the gathered villagers: 'Ishitsukuri Sama made a poor bargain. His bowl is broken and in return he has gained nothing but shame.' And they returned chuckling to their homes.

VIII

The Task of the Deputy Chief Councillor Iso-no-Kamimaro

Now the court official, Iso-no-Kamimaro, was of somewhat humbler birth than his rivals, but, for all that, he was an upright and honest young man and very serious besides. To his followers he said:

'When the Horai Mountain Swallow has built its nest, please come and inform me at once.'

His men looked at each other, and one said:

'May we know the reason for this request, your Honour?'

Iso-no-Kamimaro cleared his throat pompously, and seating himself cross-legged on his cushion, answered:

'The reason is simple. I am required by the Princess Kaguya to bring her the Cowry Shell which is inside the body of this bird.'

'Well, well!' replied his followers. 'We have killed many birds, but we have never found cowry shells inside them!'

And they laughed at such a ridiculous idea. But Iso-no-Kamimaro scolded them for their levity and bade them listen attentively.

'This cowry shell,' he said, 'is ejected from the beak of the Swallow at the precise moment that it bears an egg. But no man has ever been able to witness this act. Why? Because the Swallow takes to flight at the first sight of a man.'

'In that case, how can we even find out when the Swallow builds a nest?' queried the men and they began arguing about the seeming contradiction of the task.

Among them was a much older man, who had been many years in the service of Iso-no-Kamimaro's family, and he, seeing that his master was becoming more and more annoyed, stepped to his side.

'Sir,' he said, 'the Mountain Swallow is known to build its nest in the eaves of the Horai Mountain Temple. Many swallows nest there; but the Swallow we desire builds in the topmost roof of all. This I know from the tales told me by my mother in my childhood. Let us go there, and by the great Gate let us construct a high column from which it will be possible to look down on the Temple roof, and from where we can see without being seen. Once we locate the nest, we can easily find means to reach it.'

Iso-no-Kamimaro was delighted with this plan and

selected twenty of his men, who at once gathered building materials together and prepared for the journey to the Horai Mountain Temple. On arriving there, they worked night and day on the construction of a high, towering column with a ladder running up its length. Up and down they went, each wanting to watch from the top as the swallows wheeled and circled around them. They could see the nests and they located the topmost nest of all. But the birds were greatly astonished at these odd humans, who did not keep their two feet on the ground where they belonged, but scaled the air; and though they fluttered round chirping excitedly, not one of them would venture near its nest in the Temple roof.

'This is quite useless!' exclaimed Iso-no-Kamimaro impatiently. 'I shall never perform my task like this. Please think of something better immediately!'

His old follower reflected for a while, and finally he said:

'The trouble is, sir, there are too many of us. The birds are frightened by such a crowd. Let us all move quietly away to a safe distance and the birds will believe we have gone for good. When it is dark, let two of us creep quietly back and by means of a rope, which we will first throw over the Temple roof from the top of the column, one may haul the other up in a basket. And as we know where the topmost nest is, he can easily wait there in silence, until the Swallow lays its egg and drops the Cowry Shell from its beak. Then all the man in the basket has to do is to slip his hand quickly into the nest and pick up the shell.'

The plan appealed to Iso-no-Kamimaro and he called his men together to hear the old man's idea. But now another problem arose: all the young men had been reared in the refined atmosphere of the court, and not one was versed in country matters. Iso-no-Kamimaro

himself was equally ignorant and he said to the old man:

'How is one to know when the Swallow is about to lay an egg? None of us seems to be informed on this matter.'

'Nothing simpler,' said the old man. 'Just watch for the moment when the Swallow lifts its tail feathers and turns round seven times in the nest. That is the moment when it will settle and lay its egg.'

With this information carefully in their minds, the young men prepared to leave the vicinity of the column. But first they threw a long rope from it which caught and hung securely on the edge of the Temple roof, and both ends they allowed to drop down to the ground. Then they crept quietly to a distance and waited for night. The sun dropped low in the sky and quickly all grew dark and still. Only the sleepy twitterings of the swallows broke the silence. Meantime Iso-no-Kamimaro's followers had emptied one of the large carrying baskets and fastened it to one end of the rope hanging from the Temple roof. The young man who was to try first climbed into the basket and hid himself completely, while his companion hauled on the rope until the basket rested against the topmost roof. Here the young man inside made it steady. So they remained for what seemed a long time to Iso-no-Kamimaro and his men, who waited impatiently in the dark below.

Up in the basket the young man lay, and listened and looked attentively. After some time he heard a fluttering of wings, and peering at the nest he saw the Swallow perched on its edge. Sure enough its tail feathers were lifted and as he watched, it twitched them once, twice and then wagged them vigorously. The young man grew very excited and forgetting the rest of his instructions reached forward eagerly to feel in the nest. The bird, startled at this unexpected intrusion, flew indignantly away. The young man's hand groped all around, but to his surprise there was nothing there at all, neither egg nor shell. He had been so sure, and now his face fell as he gazed ruefully down to where his master was standing below.

Iso-no-Kamimaro was getting pains in his neck with craning to peer upward, and at last he called out impatiently:

'Well! Have you found it?'

The young man in the basket, still overcome with disappointment, could only stare with a look of stupefaction on his face and could not utter a word. Iso-no-Kamimaro, growing more impatient than ever, commanded the hauler on the rope to lower the basket to the ground at once. It was a very shamefaced young man who stepped out and hung his head before his irate master and confessed he could find neither egg nor shell.

'What a stupid fellow you are!' Iso-no-Kamimaro cried. 'You saw the bird lift its tail feathers; yet you say there was neither egg nor shell in the nest. I can see there is no other way but for me to ascend and procure the shell myself.'

So saying, he proceeded at once to take off his thick

outer kimono and to slip his arms free from the under one. The older retainer sought to restrain him, warning him of the danger of a fall, but Iso-no-Kamimaro would not listen and, climbing into the basket, demanded to be hauled aloft.

Once up on the roof Iso-no-Kamimaro lay, hardly breathing, to await the return of the Swallow to its nest. Down below his followers waited in suspense. After a while Iso-no-Kamimaro heard the flutter of the Swallow's wings as it returned and a chirping cry of 'chichi-chichi'. Peeping over the top of the basket he saw the little bird perched on the edge of the nest. Suddenly it lifted its tail feathers and flicked them vigorously.

'Ha!' thought Iso-no-Kamimaro. 'This is the moment!' and he waited breathlessly for what would happen next. The little bird turned round and round on the nest and after the seventh time crouched down with its feathers outspread. Iso-no-Kamimaro could hardly restrain his impatience, and from the way the basket trembled he could feel that his men below were sharing his tension. Flinging caution to the winds he put his head out of the basket and peered at the nest. With a whirring of wings the Swallow rose into the air and chirruped indignantly:

'Chichi-chichi! What sort of rough, rude fellow is this that interrupts me at such a moment? Look out! look out!' And as it called to the other birds they all clustered together at a safe distance from the strange form in their midst.

Iso-no-Kamimaro took no notice of their cries. He put his hand into the nest and felt all round it. Ha! what was this? Something warm and round and soft came into his hand. Exultantly he called out to his men.

'I've got it! I've got the Cowry Shell! Let me down quickly!' he cried, and clutched the object firmly in his hand.

Far below the men were equally delighted, and all
gathering together, they let the rope spin through their
hands, forgetting the danger from the sharp edge of the
Temple roof. As the rope spun past in Iso-no-Kamimaro's rapid descent, the tiles cut through it clean as a
sword. A single cry of 'Ah!' rose from the men below as
the basket came hurtling down, and Iso-no-Kamimaro
shot out of it like an arrow from a bow and fell squarely
into the open mouth of the rain-water barrel placed just
below the roof. Only his arms and legs were visible as
the splash of water subsided; then, as the men watched
motionless with horror, his head rose splutteringly from
the depths and he gave a feeble cry for help before he
sank back once more into the water. This brought them
all to their senses and with great haste they drew him out

and laid him gently on the ground. His hands were tightly clenched and the whites of his eyes shone in the darkness as they rubbed his body and called his name. After a long time his eyes closed; some of the rigidity began to leave his limbs and his breath came in big gasps. His men clustered anxiously about him and continued their rubbing and pummelling. After a little time they were rewarded by seeing his eyes slowly open.

'Oh! How my back hurts and how dark it is everywhere! But I've got the Cowry Shell. Bring candles so that we may examine it,' he said faintly and his eyes closed again.

Hurriedly the men brought candles and painfully they raised their master to a sitting position. He slowly opened his hand and all peered closer to look at the fabled shell. But what did they see? A round whitish lump, little bigger than a bean, and of an unmistakable nature. Iso-no-Kamimaro gazed at the object with dilated eyes: then with a heartfelt cry of disappointment he fell back in a dead faint.

The retainers could hardly contain their laughter, but they lost no time in preparing for the return journey. Once in their native village they called the doctor immediately. Iso-no-Kamimaro had been badly hurt in the fall, and for many weeks his followers took turns in nursing him devotedly, for each man felt himself responsible. As he gradually became better, Iso-no-Kamimaro grew more and more dejected.

'Alas! what a miserable fate! Trying to steal the Cowry Shell—its own treasure—from the Mountain Swallow. And what do I get for my pains? A lump of manure and a broken back! How everyone will laugh at me!' he sighed; and he hid his face in his sleeves and wept bitterly.

When Princess Kaguya heard of this adventure, she

was really sorry for all the trouble she had caused this honest young man, and making what amends she could, she sent him a warm letter together with many gifts and presents in a token of her esteem. This did more than anything to restore Iso-no-Kamimaro to his usual spirits and he quickly began to mend. But a sad memento of his romance remained with him for the rest of his life: from that time forth he always had a slight limp and was never able to walk without the help of a stick. Despite that, he lived to a ripe old age, but he rarely talked of the time when he tried to win the hand of the Princess Kaguya.

IX

The Emperor of the Province

After the adventures of these five young men, no one ever got so much as a glimpse of Princess Kaguya. She drew more and more into the happiness of her isolation with the old couple. The crowds ceased to come to the door and wound her with their curiosity. But despite her seclusion, the fame of her beauty spread to where the four winds of heaven touched the shores of Japan and eventually to the ears of the Emperor himself. One day he called one of his messengers to him and said:

'I hear rumours that there is a beautiful young maid called Princess Kaguya who secludes herself from the sight of men. I would much like to see her and I wish you to proceed at once to her house and escort her to my presence.'

The messenger set out immediately to fulfil this mission, and arriving at the house he knocked peremptorily on the gate. The old woman heard his knocks and came to see who it was making such a noise. When he saw her the messenger called out:

'Is this the house of the beautiful cloistered maiden?'

'Our daughter, sir, is very beautiful,' replied the old woman, 'and she does indeed prefer solitude.'

'Good!' answered the messenger. 'This is undoubtedly the maiden I seek. Kindly inform her that His Imperial Majesty the Emperor deigns to grant her the favour of a private audience and I am come to escort her to the palace.'

The old woman was delighted and at once hurried in to tell Princess Kaguya the news. When she heard it, Princess Kaguya remained unmoved. She shook her head and said:

'I am far too inferior a person to be presented to His Imperial Highness. I would not be able to lend dignity to his august palace and my coming would have an unworthy end for us all. Pray tell the messenger I do not wish to go.'

Despite all the old woman's pleadings, she remained obstinate.

When the messenger returned to the palace with the news, the Emperor was very annoyed at this affront to his kingly position, but he became all the more determined to see for himself if the rumours of Princess Kaguya's beauty were true. So he sent his messenger once again to the house, but this time he commanded him to return with Princess Kaguya's old foster-father. Needless to say, the old man had no choice but to obey. He came to the palace full of an uneasy fear, for he had heard from his wife of the Emperor's wish to see Princess Kaguya and he was only too well aware that if she had made up her mind there was no changing it. When he was brought into the audience chamber, to his great surprise the Emperor spoke kindly to him and promised to reward him with a high rank and pension if he would persuade his daughter to visit the court. Upon this, the old man became quite cheerful for he thought that now Princess Kaguya would surely agree to come out of her isolation, if only for his sake. So he returned jubilantly home.

To Princess Kaguya he said:

'I have been received by the Emperor, my daughter. He has a sincere desire to see you at court. Will you not go? It is a wonderful opportunity for you. He has long wished to have a lady for his wife and Empress. And who knows but that when he sees your gentle beauty you will be that lady? If not his wife, I am sure that he would be only too happy to make you first Court Lady. And for me, too, there is a reward: I am to be offered a high rank.'

Princess Kaguya remained silent and motionless for a long time. Finally she said:

'Father, I would do anything to see you promoted to the position you so richly deserve. But this I cannot do. Cannot, cannot do! Even if I am dragged there by force, I will refuse everything and I will only seek the opportunity to run away and die in loneliness.'

This upset the old man so much that he could hardly restrain his tears, for he could not bear to think of life for himself and his wife without their beautiful princess daughter, and he begged her not to say such dreadful things.

As the days went by, the Emperor grew more and more angry, but at the same time more and more curious, when there was no sign of the strange maiden yielding to his wishes. One evening, after a day's hunting, he found himself near the house of the old couple. He reined in his rich-caparisoned horse before the house and ordered his servant to call the old couple out. They were astounded that the Emperor himself should condescend to visit their humble house and came out to greet him,

greatly confused and embarrassed. They fell to their
knees and bowed until their heads touched the ground.
The Emperor quickly dismounted, and brushing their
salutes aside, strode directly into the house and towards
the door of the inner room. He slid open the doors without ceremony and was about to enter, when suddenly
he stepped back with a cry and flung his hands before his
face. A dazzling blaze of light had irradiated the room,
in the centre of which shone the exquisite form and
fair face of Princess Kaguya.

She sat quietly on her cushion with her small hands
folded tranquilly on her lap and her head inclined
slightly forward. Two long folds of hair hung over her
shoulders and her silk garment, now radiant in the
glowing light, spread fanwise over the floor. The
Emperor, overcome by a beauty the like of which he had
never gazed upon before, fell before her and said:

'Princess Kaguya, I am indeed the Emperor. I have
come in person to visit you, for the rumours of your beauty
fill my country. Now I know that even the rumours do
not describe what I have seen with my own eyes. With
all my heart I wish you to return with me to the palace.'

But Princess Kaguya gently shook her head.

'It is not possible. How can I leave my dear parents?
Besides, you must have guessed that I am not of this
country, and even if I came with you, I should have to
leave you one day.' And again she shook her head.

But the Emperor was not so easily rejected, and the
more he looked at her divine features, the more determined he became to win her.

'However much you refuse, I will marry you,' he
cried suddenly, his voice filled with love and passion.
He rose and was about to seize her hand when pitch
darkness fell like a pall on the room. The Emperor was
like a blind man as he groped wildly in the blackness.

'What has happened? Where are you? Where are you?' he cried.

But there was not a sound; nor could he feel so much as a thread of Princess Kaguya's dress. The Emperor sank dejectedly to his knees, but wisely realizing that there was no help for it he addressed the dark room in a pleading voice.

'Princess Kaguya, forgive me. Please forgive me. I have behaved unbecomingly. Only let me see you once more in all your beauty and I will never trouble you further.'

Hardly daring to hope she would grant his request, he bowed again to the floor.

Immediately the room was flooded with light again. He raised his eyes and before him was Princess Kaguya. So tranquil and gentle was her expression that the Emperor felt a flood of tears rising in his breast.

'Lady,' he said, 'now that I have seen you once more I can never forget you. You are more lovely than the white crests of ten thousand waves, more noble than the towering peaks of heaven, and more fair than the cascading moonlight in the valleys. Never have I seen such beauty before. Never shall I again.'

He gazed fixedly at the silent princess, then he went quickly from the room and rode back to his palace.

x

The August Moon

Four years passed by, and now it was spring. Princess Kaguya led her quiet and peaceful life with her adopted parents. But now a strange mood seemed to have taken possession of her, and often on moonlit nights she would be found seated at her window gazing at the moon as it rode in the sky. She would look so unhappy and her eyes were so filled with longing that the old couple grew more and more anxious about her.

The old woman especially was worried, for she remembered from her childhood being told that bad things happened to moongazers. She told her fears to Princess Kaguya, but the only effect of it was that the girl shut herself more closely in her own room, and for weeks on end the old couple hardly saw her.

Time went on sadly enough, and the month of August arrived. One evening, when the August moon was nearing its fulness, the old woman heard her daughter weeping bitterly in her room. She called her husband and together they went in to her.

'What is the matter with our princess daughter? Can you not tell me why you are so sad?' pleaded the old woman.

'Tell us the reason, dear Kaguya Sama. We will do anything in our power to please you,' said the old man.

Princess Kaguya struggled to control her tears as they spoke. She wiped her eyes with the border of her long sleeve, then said:

'I must tell you my secret. I cannot hide it from you any longer. I have kept silent up to now for I did not

176

wish to make you unhappy. But the time has come for
you to know all about me. You must have guessed that
I was not really born in this country and I am not of the
same race as the people here. I was born in the country
of the Moon. And there is my true home.'

At these astounding words the old couple gazed at her
in stupefaction. Then in a quavering voice the old
woman said:

'Are you . . . are you then a lady from the skies?'

Princess Kaguya smiled at her.

'Perhaps that is what you might call me, for from the
skies I have come. But now the time has come for me to
return to my own country. For many weeks the voices
from the Moon have been calling me. There is no alternative. I must go, and at the fulness of the August Moon
the heavenly messengers will come down to escort me
home. I came from my blissful land to help you for it
seemed that goodness was everywhere within you. I was
not mistaken, and it was happiness itself to be with you
all these years. Never shall I forget. Never! Never!'

Through her tears Princess Kaguya's words came
faltering and stumbling, and the old couple were beside
themselves with grief at this strange, sad news.

When Princess Kaguya's tears had ceased and she had become calmer, the old couple retired to their own room, unhappy and full of disquiet. They thought about what she had told them. The night of the full Moon was very near and they seemed powerless to do anything. But the old man grew very agitated and said defiantly to his wife that he would never allow anyone, be it angel or devil, to take their beloved daughter from them. When Princess Kaguya heard his voice, she came to her parents with a face as full of grief as the old man's, and putting her head on his knee, she said brokenly:

'I do not wish to leave you both. Only believe me. From the time you took me from the bamboo tree and carried me home in the palm of your hand, I have had nothing but love and devotion from you and I have grown to love you both dearly in return. But I belong to the Moon country. My people long for my return and to them I must go. There is no help for it.'

The house resounded with sighs and the pitiful cries of the servants who had gathered round and listened to this sad tale.

Evening by evening the Moon grew rounder and fuller and those in the house hardly slept or ate in their anxiety. At last the old man after much thought called the old woman to him and said:

'If we go on like this we shall all perish of grief. I have a good idea and I am going to get help. Keep a close watch on our daughter and don't let her out of your sight for one moment, especially at night. I shall return within a day's span.'

He left the house at once and went into the town where he hired a palanquin with the fleetest runners available and hastened to the palace.

XI

The Messengers from the Moon

When he arrived at the palace, the old man got out of the palanquin and hammered on the great gates shouting:

'On behalf of Princess Kaguya, I desire urgently to see His Imperial Majesty.'

As soon as servants brought this message to the Emperor, he hurried to the gateway where the old man greeted him.

'Your Majesty! Your Majesty! A terrible thing is about to happen to our daughter and we beg your help. Please come quickly to my house.'

At these words, the Emperor was mightily annoyed and said sternly:

'What do you mean by these strange words? Please calm yourself and explain clearly what the trouble is.'

The old man at once poured out the story of Princess Kaguya's strange birth and the even stranger story of her return to the Moon. He ended by imploring the Emperor's help before the August Moon should reach its zenith and their beloved daughter be forced to leave them.

'A miraculous tale! A miraculous tale, indeed!' said the astonished Emperor. 'But do not worry any more. I shall bring two thousand of my best retainers to your house and we shall protect the princess from these messengers whether they come from earth or heaven. Return home and keep a careful watch until we come this night.'

The old man was overjoyed and hurried home as fast as possible to prepare his wife and Princess Kaguya for the siege. First he ordered his wife to sit with the princess in the strong room where the family treasures were kept. Within it he had placed food and other necessities to last several days. Then he ordered his servants to fasten the door with the strongest and surest locks obtainable. His wife he instructed to hold fast to the hands of Princess Kaguya and on no account to lose touch with her. He himself stretched out on a mat in front of the door and prepared for his vigil.

As the evening grew dark, the Emperor and his retinue rode up. The two thousand soldiers ranged themselves along the wall surrounding the garden and on the roof of the house, where they settled like a flock of human swallows. A stillness fell over the house, and all that could be heard was the slight rustle of the men as they unslung their bows and arrows and held them in readiness to shoot.

Soon a soft radiance began to shed its light in the sky and the Moon slowly rose above them, every moment growing more full and golden till it seemed to hang above them like a bursting, over-ripe peach. The tension among the waiting men heightened, and they stiffened like crouched animals as they flexed their bows. As the shining path of the Moon widened and its rays cut into the house, the old man leapt to his feet and cried in a ringing voice:

'At the first sign of any strange thing, shoot!'

A great shout came from the men in answer to his words: 'Do not fear, old Grand-dad; even a bat will not escape us.'

Within the house Princess Kaguya sat with her hands held tightly in the hands of the old woman and sighed as she heard the brave shouts.

'Alas! However bold and courageous they are, their strength is as nothing against the might of the Moon,' she said, and she laid her head sadly on the lap of the old woman who, quite uncomprehending, held her all the tighter.

The Moon rose higher and higher and now the smallest leaves in the garden were bathed in its light. The old man shook his fist angrily at the sky.

'Whoever you are, angel or demon,' he shouted, 'I shall never let you take our daughter away from us! I shall tear out your eyes with my long nails! I shall kill you, whoever you are!'

But not a sound came from the sky. Only, as midnight drew near, an unearthly brilliance, ten times brighter than ten thousand moons, filled the sky and the watchers cowered back in terror, almost blinded. From the highest point in the heavens, a wreath of white clouds slowly massed together and gently and silently began to drift downwards. As it approached, the tensed watchers could see that many beautiful creatures seemed to be grouped upon it; some standing, some sitting, and all dressed in shining kimonos of rainbow-light colours. They were so numerous that it was impossible to say if there were a hundred, two hundred, or many thousands. In the spellbound silence they drifted to a point just above the roof top and there rested quietly.

The waiting soldiers fell to trembling so much that no matter how they tried, they could not stiffen their limbs to aim their arrows. Then one of their number, wiser than the rest, said as he fell to his knees:

'Pray, all of you! Only prayers can help us now!'

To some a moment's strength seemed to be given and several bows twanged fiercely as their arrows cleft the air. But all fell wide and the heavenly visitors remained motionless and tranquil on their floating white platform. Before anyone could realize how or from where it had come, the watchers saw a graceful palanquin gliding down the broad path of the Moon's light. It was guided and held by a host of beautiful beings led by one more gracious and lovely than all the rest. He seemed to be a person of great importance, and as he approached, he called out in a clear, sweet voice:

'Miyatsukomaro San! Please come before us.'

At the strange power of this lovely voice, that sounded like the waters of a meandering stream, and in surprise at being addressed by his family name, the old man climbed fearfully on to the roof. At sight of the celestial being, all his anger left him and he could only kneel in humble obeisance and listen as the voice continued:

'You have done well in caring so lovingly for one who is our dearest treasure. And in doing so you have become rich and happy. Have you not, Miyatsukomaro San?'

'Indeed! Indeed!' cried the old man. 'And we are deeply grateful.'

'Now it is time for her to return to her own people. You have been parents to her for many years. Yet those years are as a minute in the country of Princess Kaguya. There her people await her, for there she belongs. There is nothing you can do now to keep her longer with you. Miyatsukomaro San, please release her at once.'

But still the old man could not bring himself to do so, and casting round in his mind for some excuse, he said:

'My noble Lord, is it possible that you have mistaken our daughter for some other celestial girl? You speak of her being with us only a short while, but indeed she has been our child for twenty years. Surely you are mistaken?' In a last desperate plea, he added: 'Anyway, she is very ill just now and cannot come out from the house.'

But there was no answer from the young Moon God, and all were silent as he guided the palanquin down to a resting place on the roof. Again in that clear and liquid voice he spoke:

'Princess Kaguya! Princess Kaguya! We have come for you. Leave this house and come to us! Your people await you and now you must return to them. Come, your palanquin is waiting.'

All listened silently. No one tried to move. And in the inner room the old woman sank down in fear as a sudden numbing paralysis seized her limbs. Weakly her hands fell from their close hold on the girl, and her eyes dilated in awe as the door slid open without a sign or sound of human agency, though it had been locked and barred. Her cry brought the old man into the house and he too sank down to the floor in fear, as he saw the lovely form of Princess Kaguya gliding unchecked through the open door. She looked on them both with a face of the utmost tenderness and laid her hand gently on the old man's shoulder as she said:

'Do not weep, I beg of you. Only remember always our happy years together. But now there is no help for it; I must leave you to return to my country and my own people. Will you not lift your faces to me and come and wave me farewell on my journey?'

But the old man shook his head.

'Dear child, dear child,' he murmured, 'we have been your father and mother for all these years; now we are grown old and it is now we need you most. How can you leave us? Take us with you, I beg you, to your home in the sky, for without you we have no wish to live.' And he and his wife wept bitterly together.

Princess Kaguya was greatly troubled by their grief and could hardly restrain her own tears; but it was at this moment that one of the messengers from the Moon entered the house. He bore in his hands a precious ornamented box, which he placed ceremoniously before the princess. From it he took a shining pitcher and a flowing robe of the most delicate gossamer silk in colours that irradiated the room. It seemed to be made of a thousand beads of rain. To Princess Kaguya he said:

'This is the robe called "Hagoromo". When you put it on, it will wipe away all touch of human dross. And in this pitcher is the potion that will bring you forgetfulness of all your present sadness.'

Princess Kaguya turned aside and through her tears begged him for a moment's respite. She took off her beautiful outer kimono and kneeling before the old couple said:

'Keep this in memory of me. Where it is, there shall I be too. Please now, dry your tears. Do not be unhappy, for my sake.'

She touched them reverently on the shoulders. Then, taking one of her writing brushes from a lacquered box, she wrote a poem of farewell on a scroll. She rolled it and handed it with the pitcher to her old father and said:

'Give this poem and this pitcher, father, to the Emperor. My poem is my farewell to a beloved man. The pitcher holds a potion that, when he tastes it, will bring him eternal youth.'

With a last longing and loving look at the old couple, she turned to the messenger and said:

'I am ready to return with you.'

She stood silent and composed as he draped the robe about her shoulders. A look of majesty fell upon her. Joy and divine happiness flowed into her cheeks and an ethereal air surrounded her whole being. In this one instant she became so far distant and forgetful of her human life that to the old couple she was already a stranger. But they still could not believe that their darling child was lost to them and they followed her out with frenzied, broken cries as she put her hand into that of the young god of the Moon and stepped with him into the shining palanquin. The great host of messengers rose in the air. As they circled the palanquin, they broke into a flood of rejoicing song, and a path of moonlight opened before the cavalcade. Deaf to the frantic cries of her broken-hearted foster-parents, Princess Kaguya floated joyously upward to her home surrounded by the shining host, until not a point of light from their robes was visible to the straining eyes below.

XII

The Emperor's Pilgrimage to Mount Fuji

When the Emperor received Princess Kaguya's poem and the potion, he sighed deeply and fell into a long meditation. His heart was torn asunder, for the full depth of his love for the beautiful Princess Kaguya suddenly came upon him.

'Ah! Since the princess is no longer on this earth, I have no wish to prolong my life,' he murmured.

He withdrew into himself. For days on end he refused to speak to anyone, and he carried with him everywhere his spirit of desolation. He shunned his retainers and turned away from all the entertainments and diversions they planned for him.

One day he called them together and said:

'Which is the mountain nearest to heaven in our country?'

'Why, sir, it is Mount Fuji,' they replied in some surprise, for this was well known to everyone.

'Good!' said the Emperor. 'Make preparations for a journey. We shall set out for Mount Fuji immediately. In its fiery depths I shall burn this poem and the potion which day and night consume my heart with memories of the lost Princess Kaguya.'

They set out and arrived at the foot of the divine mountain. They climbed throughout the night and reached the top as the rising sun exploded through fields of clouds on the horizon. With his noble retainers ranged about him, the Emperor committed Princess Kaguya's poem and the pitcher to the red glowing crater. As they burned the smoke rose thick and black. Gradually it cleared and left a fine wavering thread of smoke which spiralled up and up towards the heavens.

'It is reaching for the country of Princess Kaguya,' murmured the Emperor.

With these words, he left the summit and began the long journey home. But many times he looked back and always there was the thin thread of smoke ascending into the still air. And so there is to this day.

The Tongue-cut Sparrow

IT was autumn, and the dawn was breaking. The forest was afire with the red of the maple trees; the cranes glided down to the watery rice-fields to dab for their morning meal; the croaks of the bull-frogs rumbled from the river banks; and Mount Fuji, wreathed in clouds, breathed idly and contentedly on the distant skyline. It was a season and a morning dear to the old woodcutter's heart, and neither his poverty nor the sharp tongue of his irascible wife disturbed his tranquillity and happiness as slowly, with bent back and grasping a stout staff in his hand, he tramped through the forest to cut the day's fuel.

The birds knew him as a loving and gentle friend and chirruped in time to his walk, or flew from branch to branch along his path, waiting for him to scatter the millet grains which he always carried for them in a small bag tucked in his kimono sash. He had just stopped to throw the millet on the ground, when above the twittering he heard a plaintive cry of 'Chi! chi! chi! Chi! chi! chi!' It seemed to come from a nearby bush though there was nothing to be seen. The woodcutter, sensing that a bird was in distress, went quickly to where the cry appeared to come from, and parting the bush, saw a small sparrow lying in the grass panting with fright and unable to move. Picking it up gently in both hands, he examined it and found that one of its legs was wounded. He tucked the sparrow into his kimono against the warmth of his breast and returned home at once to attend to the little sick creature.

His wife stormed bitterly at him when she learned the reason for his return and showered ill-natured complaints on him at the prospect of another mouth to feed, even though it was such a small one. The woodcutter, long resigned to her harsh tongue, went about quietly and unconcernedly tending to the sparrow. He laid it on an old cloth in a corner and fed it with warm rice-water and soft grains of millet. Day after day he cared for the little bird and with such unfailing devotion that, when the first snows came, its leg had already mended and its body was well and strong.

While it was ill the sparrow rarely ventured from the cage the woodcutter had fashioned for it, but as it became stronger, it became more venturesome. It took to hopping about the straw mat room and the wooden veranda outside, but ever with a watchful eye on the woodcutter's wife, who loathed it and lost no opportunity for attacking it with her broom and heaping on its head the wrath of the seven gods of thunder. With the woodcutter it was different. The sparrow adored his gentle rescuer and the old woodcutter in turn loved the sparrow with all the warmth of his tender heart. Each evening it perched on the thatched roof to await his return from the forest. As he emerged from the darkening trees, it would set up an excited welcoming cry of 'Chun, chun, chun!' and fly round his head, sit on his shoulder, and pour its twitterings into his ear.

In the mornings it was a different story. As soon as the sparrow saw the old man preparing to leave, it huddled forlornly in the corner of its cage and sang its plaintive 'Chi! chi! chi! Chi! chi! chi!' The woodcutter, equally sad at parting from his pet, would take the little bird gently in his hands, and stroking the soft feathers, say:

'Well, well, now! Do you think I am leaving you for ever? Content yourself, my friend. I shall be back before the last light leaves the trees.'

One morning the old man went off as usual, having first told his wife to take good care of the sparrow and give it something to eat during the day. The old woman merely grunted, muttered a curse, and proceeded with her preparations for washing out their spring kimonos. She drew water from the well and filled the great wooden pail, and in this she placed the fine cotton kimonos to steep. Then the long bamboo poles had to be wiped clean and slung from branch to branch of the trees. On these the kimonos would be threaded from sleeve to sleeve, so that they would dry quickly in the light breeze that fanned the trees. Next she put some of her precious store of rice-flour into a deep earthenware bowl and mixed it with a little of the water into a glistening white paste. Today she took especial care to mix it fine and smooth, for she was preparing her own and her husband's best kimonos for the ceremonious advent of spring, and it was her custom to soak them in the rice-paste to give them a fine glossy sheen. Their supply of food was scanty enough, but she always managed to save enough of the flour for this yearly ritual.

Leaving the bowl of paste on the veranda, she squatted down by the wooden tub and began the long task of rubbing and steeping, steeping and rubbing, until the kimonos were clean and fresh as young bamboo shoots.

It was long past midday before she finished, and the poor sparrow, now ravenous, was singing its best to win the old woman's heart and her millet grains. But to no avail. She continued with her washing as if the bird did not exist, and the sour lines on her face told it that she had no intention of giving it anything. Dejected, it flew to the veranda, and seeing the bowl, perched on its rim. Whatever the white paste was inside, it looked good, smelt good, and, 'It tastes delicious, chun! chun!' cried the sparrow as it withdrew its beak and the rich rice-paste passed over its tongue.

'Oh! Oh! Oh! What a dish! What a find!' it chirruped in delight, and down went its beak again and did not reappear until the bottom of the bowl gleamed bare and clear in the midday winter sun. The sparrow hopped from the bowl on to the veranda and was preening itself in the sunshine when the old woman returned with the kimonos to dip them in the paste. When she saw the empty bowl, her whole body began shaking with hatred and anger, and seizing the sparrow before it had time to dodge out of her reach, she yelled:

'You did it! You did it! You gluttonous, grasping scavenger! Now I'll put an end to that pretty song of yours for good. Do you hear? For good! For good!'

As her voice rose to a screech, she pulled a pair of scissors from her pocket and forcing the sparrow's beak apart, slit its tongue with the sharp blades and flung the poor creature to the ground. The sparrow turned and churned the dust and its wings beat the earth in agony. Cries of pain formed in its throat, but no sounds came from its beak. Several times it tried to lift itself from the earth, but its sufferings seemed to anchor it. Round and round it struggled and fluttered. Then, with one last effort of its little pain-filled body, it rose in the air and disappeared over the tree-tops of the forest.

Returning home that evening, the woodcutter was greatly surprised not to hear his usual welcome as he approached the hut. His pet was nowhere to be seen. And no glad 'Chun, chun, chun!' broke the evening stillness. Perturbed and uneasy, he went straight to its cage but found it empty. Turning to his wife he asked:

'Where is our little Chunko?'

'The nasty creature ate every morsel of my rice-paste: so I slit its tongue and drove it away. Wherever it is now, it is better than being here; for I could stand the wretch no longer,' his wife replied in anger.

'Oh, how pitiful! How pitiful!' cried the woodcutter in anguish, as if his own tongue had suffered the fate of his little sparrow. 'What a cruel, what a wicked thing to do! You will suffer for this evil, indeed! Where is my little friend now? Where can it have gone?'

'The farther the better for my part,' snapped back his wife, untouched by her husband's distress. 'And a good riddance into the bargain!'

That night the woodcutter could not sleep. He turned and tossed in wakeful anxiety for his little bird, calling out from time to time in the hope that it might answer. When at last light came, he rose and dressed quickly and went to the forest in search of it. For a long time he wandered calling out:

'Tongue-cut sparrow, where are you? Where are you? Come to me, my little Chunko!'

But only the croaks of the bull-frogs, the cries of the cranes high overhead, and the chirrupings of the forest birds answered; the gay, glad song of 'Chun, chun, chun!' was nowhere to be heard. All the morning he searched and far into the afternoon, forgetful of food or weariness and with thought only for his little friend. As the evening light settled over the forest, turning the shadowy trees to the shapes of menacing giants and ferocious beasts, he sat down at the foot of a tree, exhausted and desolate, but still calling out:

'My little tongue-cut Chunko, where are you? Where are you?'

Overcome by the sadness in the woodcutter's voice, some sparrows, perched above him in the tree-tops, flew down to greet and to talk to him. The old man was over-joyed to see them and begged them for news of his friend. The birds were deeply moved by the woodcutter's grief, and twittering among themselves, they finally said:

'Grandpapa San, we know your Chunko well and where it lives. Follow us and we shall lead you to its home.'

The woodcutter, all thought of his weariness gone, sprang up and started out after the sparrows. He followed in the darkness for a long time, till at last they came to a clearing and there, in the midst of a moss-covered patch, surrounded by bamboo saplings, was a house gaily lit with lanterns hanging from the thatched eaves. Immediately a throng of sparrows came out to welcome him. They lined up before him and bowed deeply until their beaks touched the ground. They showed him into the house with every courtesy, helping him to remove his straw-bound clogs and putting soft slippers on his feet. They led him along a corridor of shining cedar wood to a room of newly-laid straw matting. Here he courteously knocked off the soft slippers and entered in his cloth socks. The sparrows pulled back the decorated sliding screens of an inner room to reveal little Chunko surrounded by a flock of attendants, sitting on the floor awaiting his arrival.

'Oh! little friend, I have found you at last! I have looked in every tree in the forest to bring you back and comfort you and ask your forgiveness for the wickedness of my wife. And your tongue? Is it healed? How I grieved for you! I am overjoyed to see you again,' the woodcutter cried with the tears trickling down his cheeks.

'Thank you, thank you, Grandpapa! I am completely healed. Thank you! I, too, am overjoyed to see you,' wept the little sparrow and flew to the shoulder of the old man, who stroked it gently and tenderly.

'But, now, you must meet my parents,' said Chunko.

So saying, the sparrow led him into another room and presented him to its parents, who knew already of their child's rescue from death and the great kindness bestowed upon it during the long days of its illness by the old woodcutter. Bowing low, the parent birds expressed their grateful thanks to the old man, murmuring with deep gratitude that their obligation to him could never be repaid. They summoned the serving birds and instructed them to prepare a feast. As an honoured guest, they seated the old man nearest to the alcove in which a silk scroll inscribed with a poem hung. The old woodcutter was lost in wonder at the great beauty of the table and its furnishings. The chopsticks were of pure ivory, the soup-bowls of gilded lacquer, and the serving dishes were from the finest kilns in the land. Exquisite dish followed exquisite dish and all was served with delicacy and taste.

After the feast a group of elegant and gaily-kimonoed young sparrows entered, and to the accompaniment of two older birds—one who plucked the strings of the samisen and the other who chanted the words of the song—they performed the famous classical dance, 'The Wind among the Bamboo Leaves.' At that moment a light wind rose in the bamboo grove outside, shaking the branches and rustling the leaves in harmony with the sweet voices of the dancers as they joined in the words of the song. As the dance finished and the wind among the leaves died away, the dancers bowed gracefully before disappearing into the inner room. Almost immediately they were followed by a second group, all carrying many-coloured paper parasols. The music of the sami-sen became sparkling and gay; the parasols twirled and spun; the dancers' feet beat 'tom, tom, tom'; and the lanterns hanging from the eaves swayed in rhythm with the dance. The woodcutter's eyes sparkled, he beat time with his chopsticks, and he was lost to all but the merri-ment of the wonderful scene.

The music faded and the dancers bowed and pattered out. Thoughts of his wife began to trouble the old man and reluctantly he told his hosts that he must return home. The sparrows were deeply disappointed and tried hard to dissuade him, but the woodcutter said that it would be unkind to leave his wife alone any longer and that he must return. Never before had he known that life could be so good, so gay, and so gracious; never would he forget this evening and the rare kindness of his honour-able hosts. But now he must leave. They pressed him no further. Then the father bird spoke:

'Honourable and gentle woodcutter, we are deeply conscious of your greatness of heart and the loving care you bestowed upon our only child. You came to love Chunko as your own, and Chunko loved you as a father. We want you to remember that our humble home will always be yours, our unworthy food will be your food, and all we possess we shall always share with you. But tonight we wish you to accept a gift from us as a token of our unbounded gratitude.'

At this, two wicker baskets were brought before the old man by the serving birds and placed on the floor.

'Here are two baskets,' continued the father bird: 'one is large and heavy; the other is small and light. Which-ever you choose, my honourable friend, is yours, and is given with the heartfelt wishes of us all.'

The woodcutter was deeply moved and tears filled his eyes. He looked at the parent birds for a long time unable to speak. At last he said:

'I have no wish for many possessions in this world. I am old and frail and my time on earth will not be much longer. My needs are very small. So I shall accept most gratefully the smaller basket.'

The serving birds carried the basket to the entrance hall and there they tied it on the old man's back and helped him on with his clogs. All the sparrows gathered at the door to wish him farewell.

'Goodbye, my little friends. Goodbye, little Chunko! Look after yourself! It was a wonderful evening and I shall never forget it,' said the old man and bowed courteously many times. With a final wave of his hand, he left the grove and disappeared into the blackness of the forest with a flock of sparrows flying in front to put him on his way.

When he reached home the clouds were already glowing with the morning sun. He found his wife as angry as a November storm because of his long absence and her fury was unleashed over the poor woodcutter's head. Suddenly catching sight of the basket on his back, her tirade stopped.

'What's that you've got on your back?' she said in a voice filled with curiosity.

'It is a gift from the parents of little Chunko,' replied her husband.

'Well, why do you stupidly stand there and not tell me? What is it? What have the creatures given you? Don't stand there like someone dead! Off with it from your back and see what's inside!' carped her greedy voice, and grasping the straps, she trailed the basket from his shoulders and tore open the lid.

A burst of dazzling brightness momentarily blinded

her avaricious eyes, for inside lay kimonos soft as the morning dew and dyed with the petals of wild flowers, rolls of silk spun from the plumes of cranes, branches of coral from the seas of heaven, and ornaments sparkling brighter than the eyes of lovers. They both gazed in silence, dazed and bewildered; these were riches beyond even the world of their imagination. 'A poet's dreamings,' murmured the old man, and fell into silence again. The old woman drove in her hands and let the ornaments trickle through her trembling fingers.

'We are rich! We are rich! We are rich!' she repeated over and over.

Later that day the old man recounted the story of his adventure from the beginning. When his wife heard that he had chosen the small basket when he might have had the larger one, she burst out in anger.

'What sort of a stupid husband have I? You bring home a small basket when with a little more trouble you could have brought home twice the quantity of treasures. We would have been doubly rich. This very day I will go myself and pay the birds a visit. I shall not be so senseless as you. I will see to it that I return with the big basket.'

The old woodcutter argued and pleaded with her to be content with what they had. They were rich beyond the wealth of kings—enough for them and all the generations of their relatives. But her ears were stopped by the thoughts of her clutching, covetous mind, and grasping her outer wrap, she rushed out in a fever of anticipation.

As she had a good idea of the whereabouts of the sparrow's house from her husband's description, she reached the bamboo grove before midday.

'Tongue-cut sparrow, where are you? Where are you, little Chunko? Come to me!' she cried.

But her voice was harsh and even her smooth pleadings could not conceal the cantankerousness of her nature. It was a long time before any bird appeared. At last two sparrows flew from the house and curtly asked her what was her business.

'I have come to see my friend, little Chunko,' she answered.

Without saying another word, the sparrows led her to the house, where she was met by the serving birds, who, also quiet and reserved, led her along the corridor to the inner room. She was in so much hurry that she refused to stop to remove her wooden clogs and the sparrows were horrified at such insolent bad manners. When little Chunko saw her, it flew terrified to a roofbeam.

'Ah! I see that you are quite recovered, my little pet. I knew I had not really hurt you!' she said in a honeyed voice. Then forgetting all womanly modesty and oblivious of the cold atmosphere about her, she blurted out:

'I am in a hurry. Please do not bother to dance for me. And I have no time to eat anything either. But I have come a long way, so please give me a souvenir of my visit quickly, as I must return at once.'

In silence the serving birds brought in two baskets, one large and heavy and one small and light, and placed them before her.

'As a parting gift from us, please accept one of these baskets,' said the father bird. 'As you see, one is large and heavy and the other is small and light. Whichever you choose is yours.'

Barely waiting for the parent bird to finish speaking, the old woman pointed eagerly to the large basket.

'It is yours,' said the bird gravely.

In the hall, with many shoves and heaves, the sparrows hoisted the basket on to the old woman's back and

bowed her in silence out of the door. She wasted no time in bows in return but hastened off into the cover of the forest, staggering under the weight of the basket.

No sooner was she out of sight of the bamboo grove than she dragged the basket from her back and flung open the lid. Horrified she fell back as monsters and devils poured out with eyes shooting flames, mouths belching smoke, and ears emitting sulphurous clouds. Some had seven horned heads that lolloped and rolled on their slithery bodies, some had arms that writhed and coiled like snakes waving and searching blindly through the sulphurous air. Bodies, tenuous and billowing and spiked with the horns of great sea-shells, floated upwards and outwards; among them one in the semblance of a young girl with floating black hair whose sole feature was a single eyeball set in the centre of a blank, white face. All these rose and bent and drifted over the horror-stricken body of the old woman.

'Where is this grasping, greedy, wicked woman?' they screamed, and the snaky arms groped and twisted round her. Suddenly all the monsters shrieked with one searing, ear-splitting voice:

'There she is! There is the evil-minded hag! Let us blow sulphur in her eyes and they'll be greedy no longer. Let us embrace her to our shell-spiked breasts and destroy the wickedness in her flesh. Let us peck and nibble her with our forked tongues until she dies, dies, dies.'

Panic-stricken the old woman fled, all feeling frozen out of her body. Through bush and bramble and water she sped with the swiftness of the wind, the monsters in mad pursuit behind.

'Peck her, nibble her, blow sulphur in her eyes; puncture her flesh with our spiked breasts,' they screeched.

'Oh! Buddha! Help me! Save me from these devils!' the old woman screamed.

Their bodies floated over her, their blindly groping arms stretched out to enfold her. Suddenly there was a burst of light among the trees. It was the setting sun showering the sky with rose and gold. As the golden radiance flooded the forest the monsters huddled back with yells of dismay, and turning in panic, they vanished into the darkness of the trees and were seen no more.

The old woman stopped, breathless and trembling, her body sick in every pore. The radiance in the forest was now dying, and dreading the monsters' return, she started off again, exhausted and trembling at every step.

When she reached home, her husband, shocked at her pitiful state, ran out and helped her to the veranda, where she sat panting for some time before she was able to speak.

'What has happened to you? What has happened to you? Do please tell me!' pleaded the old man.

His wife, after telling him her story, said:

'I have been ill-natured, evil-minded, and greedy all my life. This is the retribution I have deserved. I have had my lesson, a bitter one, but not perhaps so bitter as the life I have led you. Now I know how evil I have been. From this hour onwards, I will mend my ways. I will try to be a kinder, gentler woman and a better wife to you, my dear husband.'

He placed his hand on her shoulder and they both knew that the bad days were gone for ever. For the years that were left to them they knew no want and never a harsh word passed the old woman's lips. The sparrows became their closest friends and each paid regular visits to the other's home. Long after the old couple died, the sparrows commemorated the story of the old man and the old woman in a song, and for all I know they sing it to their children still.

The Lucky Tea-kettle

ONE morning, in the days when Mount Fuji was worshipped as a god and as the most divine of all nature's children, a young badger, full of gaiety and the warmth of the first sun of spring, gambolled on a remote moorland with abandon. He rolled over and over, turned somersaults, skipped with the long stems of wild hare-bells, and squeaked with delight as he dodged the gorse bushes. He stopped at intervals to beat his stomach like a drum with both furry paws—an endearing trick badgers have when they are happy. It makes a jolly 'pon-poko-pon-pon' sound, and if there are children about they run to join in the badgers' playtime antics.

He was careless of all but his own happiness, and jumping wildly into a clump of tangled grass, he failed to see a straw rope with a noose at the end which dangled from a bamboo stake. The noose slipped down over his shoulders and held him fast. In terror he rushed to escape, but the noose only became tighter, and the more he struggled the tighter it became.

'Oi! Oi!' he screamed out. 'Oi! Oi!'

His cries reached the ears of a tinker, who was at that moment trudging home over the moorland. Quickly he slipped his large bamboo basket from his shoulders and ran to the spot.

'Oya! a poor little badger caught in a trap!' the tinker cried in surprise, and at once he set the creature free.

'Now, Badger Chan, you run home before you are caught in another wicked trap,' advised the tinker firmly but kindly. He stroked the badger's ruffled fur where the noose had bound it, and giving him a few affectionate pats, said again: 'Now, be off with you.'

The badger was overcome with the tinker's kindness and burst into tears of gratitude.

'How can I ever repay you?' he wept.

'By returning safely to your home at once,' replied the tinker, and stroking the little badger again, he started off on his way.

The badger stood for some time watching him go and wondering what he might do to help his rescuer. Suddenly an idea struck him, and calling on his one magic power, he started to transform himself into a beautifully ornamented tea-kettle. His body grew fatter and rounder and his fur sleeked into the rich brown lustre of an antique kettle. His tail curved into a handle and his four furry paws shrank to become the four feet. Only his pointed and bewhiskered nose projected where the spout should be. Taking advantage of a pause in the tinker's stride as he stopped to adjust the basket on his back, the badger tea-kettle hopped nimbly into it and the unsuspecting tinker continued on his way.

'I have returned, wife,' called the tinker when he reached home.

His wife came running and bowed a greeting as he lowered his bamboo basket on to the wooden veranda in front of their small hut. As he removed his straw sandals she caught sight of the tea-kettle.

'Ara! Ara! What is this?' she cried, and she and her husband looked at the kettle in astonishment.

They carried it to their room and placed it on the floor, where its dull sheen glowed against the poor threadbare straw matting. They knelt beside it and gazed at it in silent admiration.

'It is indeed a miracle, a miracle,' murmured the tinker.

'There is no more beautiful kettle in the whole of Japan,' murmured his wife in reply. 'Where did you find it?'

'I do not know where it came from,' answered the tinker. 'Before this moment I have never set eyes on it.'

They fell into silence again, their eyes lost to all but the delicate form of the little kettle.

'It is exquisite enough to give as an offering to the Morin temple,' thought the tinker. And then aloud, 'What do you say, wife, shall we offer it to the Morin temple?'

'It is too good for us, and I know that the priest will be happy to receive such a treasure,' replied his wife.

The tinker picked the kettle up carefully from the floor, wrapped it in a cloth, and started out for the temple. When the priest saw the kettle he was greatly surprised, for he could see at once that it was a valuable treasure and wondered how one so poor as the tinker had come by it. He was even more surprised by the tinker's story, and as there was no way of finding out who the owner might be, he readily accepted it for service in the ancient Tea Ceremony at the temple.

When the tinker had departed, the priest examined the kettle more closely and thought to himself: 'It is indeed a kettle of exquisite rarity. I shall invite some friends and have a viewing party.'

Filled with curiosity as to what the new temple treasure might be, the friends arrived. They sat in a circle on large square cushions on the floor. The paper-screened doors were slid back to their fullest extent and the room at once became part of the garden with its carefully laid stepping-stones, its large stone lantern and dwarf pine trees. It was a day perfect for treasure viewing.

After the first cups of green tea had been served, the priest brought a fine silk cloth and spread it on the floor. On it he placed the kettle and the guests fell to examining it and praising its simplicity of line, its symmetry, and the lustre of the metal. They were exceedingly curious to know where the priest had acquired it and they listened with rapt attention as he recounted the tinker's story.

'A truly beautiful kettle and worthy of being used for the Tea Ceremony of the temple,' said the guests.

'Indeed! Indeed!' answered the priest. 'This evening I shall perform the Tea Ceremony and use it. It will add to the purity and refinement of our ritual. Come this evening, my friends, two hours before the sun sets, and we shall partake in a Tea Ceremony party.'

That evening, two hours before sunset, the friends gathered in the small outer guest hut in the garden. The priest filled the kettle with water and placed it on the low charcoal brazier. He was about to lay out the Tea Ceremony utensils in their prescribed order, when he heard a loud cry, 'Too hot! Too hot!' and to the amazement of everyone there, the kettle rolled off the fire with a bump and a splash to the floor. Out from it sprouted the pointed nose, the fluffy tail and the furry paws of the badger. He skipped and hopped gingerly round the room leaving a trail of steam behind him and all the while shouting, 'Too hot! Too hot!'

The priest fell back in fear and shrieked: 'It is a ghost! It is bewitched!' and fled from the room with his guests close behind him. His young apprentices heard the cries and came rushing in with brooms and dusters to defend him, shouting, 'Where is the ghost? What has it done to you, Father?'

Trembling, the priest and his guests put their noses round the door and looked in fear at the kettle, which had now resumed its shape and was reposing innocently in the corner. Pointing a shaking finger at it, the priest said:

'I put that kettle on the fire to heat the water and suddenly it jumped off crying, "Too hot! Too hot!" and came leaping round the room.'

The apprentices chattered among themselves over this miraculous event and gingerly prodded the kettle with their brooms and long-handled dusters. One of them fetched a stone pestle and with this he prodded the sides of the kettle, saying, 'Come, phantom! Show your horns and cloven hoof!'

But nothing happened and the kettle remained immobile and innocent as before.

The priest, however, had suffered such a rude shock that he decided to return the kettle to the tinker. Accordingly he sent for him and, after explaining all that had happened, begged the tinker to take the kettle away with him again.

'Well, well, this is certainly a very remarkable kettle,' said the tinker and he wrapped it carefully in the cloth and returned home with it.

That night, after his wife had spread their sleeping mats out on the floor, the tinker placed the kettle at the side of his pillow and the couple retired to rest. During the night the tinker was awakened by a voice saying:

'Tinker San, Tinker San, wake up!'

Sleepily rubbing his eyes he saw to his amazement that the kettle had sprouted the sharp, whiskered face, fluffy tail, and furry paws of his little badger friend.

'I was so grateful to you for your timely rescue,' the little creature said, 'that I determined to help you in some way. So I changed myself into a tea-kettle and hid myself in your basket. I thought you would probably sell me and obtain at least some temporary ease of your poverty. But your nature proved more unselfish even than I had dreamed of, and you and your wife thought only to hand me over to the temple priest. But my interest was in helping you; so I devised a trick to scare the priest into handing me back to you again.'

The little bewhiskered kettle chuckled as he continued: 'One day I hope to end my days in the safe shelter of a temple; but meanwhile I assure you there is much we can do together. Now I ask you to open a show-booth and I will perform for you and make your fortune. I am really quite a skilful fellow, I promise you!'

With this the badger tea-kettle fell to performing such amusing dances and acrobatic antics that the tinker was enchanted and saw that there were indeed great possibilities in what the badger said.

The very next day he set about putting up a show-booth and outside it on long, streaming banners he advertised:

The Living Tea-kettle!
The only Live Tea-kettle that
Dances and Walks the Tight-rope!

The news spread with the swiftness of the wind across the country-side, and large crowds flocked from near and far to gaze at the streaming banners and the bright coloured curtains of the booth. The tinker sat on a high stool at the entrance and called out:

'Welcome! Welcome, honourable people! Your only chance to see a living tea-kettle! It dances with the grace of a trembling bamboo leaf! Welcome! Welcome, honourable people!'

And he and his wife could hardly keep pace with selling tickets as the people pushed and shoved to gain admittance.

Inside the booth, the air was tense with expectancy. The young girls in their bright coloured kimonos, and young ladies with their hair piled high in rolls glittering with ornaments, twittered like a flock of starlings. Mothers with their babies strapped on their backs chattered ceaselessly to anyone who would listen to them. And the farmers in their conical rice-straw hats gabbled no less than their wives. It was a sea of colour and babble and the only topic of talk was the miracle of the living tea-kettle.

Kachi-kachi-kachi! The ringing sound of the wooden clappers heralding the start of the show stilled the excited chatter. The audience was tense with anticipation as the curtain rolled back to reveal the tinker kneeling in the centre of the small stage. He was dressed in a fine new kimono for the occasion, and bowed low to the audience in greeting. At this moment the badger tea-kettle came running on to the stage, and amazement rippled through the audience as he squatted down at the side of the tinker and bowed deeply with the grace and charm of any lady. A whispering like the rustle of dry rice stalks broke out among the spectators:

'Look! Look! The tea-kettle is bowing to us!'

The tinker quietened his delighted patrons with a gesture, and in the loud, dignified voice of a showman, announced:

'Honourable people! This rare and wonderful Only Living Tea-Kettle will dance.'

With that the badger tea-kettle opened a small fan and performed an old Japanese children's dance to the delight of the spectators. When the dance was finished, the tinker had to shout through the storm of applause to announce in the biggest voice possible:

'AND NOW HONOURABLE PEOPLE! THE CHIEF ATTRACTION OF THE EVENING! THE WORLD'S ONLY LIVING TEA-KETTLE WILL WALK THE TIGHT-ROPE!'

The badger tea-kettle then bound a cotton scarf round his head as a sign that he was ready to perform an important and dangerous act. The tinker lifted him up on to a rope which stretched high across the stage and handed up to him a paper parasol and a fan. The badger tea-kettle then performed such spectacular tricks and antics on the tight-rope that the crowd yelled with delight and approval and stamped their feet and clapped their hands in tremendous applause.

The Tea-Kettle became famous and every day folk flocked from town and village, from mountain and moorland, to see him perform; and the tinker and his wife quickly became rich beyond their wildest dreams.

One day the tinker, who had grown fonder and fonder of his little friend, said:

'My dear little colleague! Already you have done far too much for us, and I fear you are becoming tired and overworked on our behalf. I assure you we now have more than we need, and though we shall grieve deeply at parting from you, yet we wish you to return to your own form, whichever it is, and to lead your own life in the way you desire.'

He closed the show-booth up from that day and no more shows were ever held. The badger tea-kettle, who in fact really had become very tired, was overjoyed at the success of his plan to help the kind tinker and now wished for nothing better than to end his days in the quiet peacefulness of the temple. So he said farewell with many deep and affectionate bows and salutations to his human friends and returned to his tea-kettle form. Very carefully the tinker carried his dear little partner to the Morin temple, and there he related to the priest in detail all that had happened to him since his last visit. The good priest was full of remorse that he had so much misjudged the badger tea-kettle, but was delighted to hear of the good fortune that had befallen the kind tinker.

'Certainly this is a rare and valuable tea-kettle,' said the priest, 'and never again shall I place it on the hot charcoal of the brazier.'

He carefully put it in a place of honour in the temple, where it remained for many a long day and may be there still for all I know.

Glossary

Chan	Informal form of 'San' only used for children, intimate friends and family members.
Mochi-cakes	Cakes made from pounded rice.
Sake	Wine made from rice.
Sama	A more polite form of 'San'.
Samisen	A three-stringed instrument played with a wooden or ivory plectrum.
San	Mr, Mrs, or Miss.
Sayonara	Goodbye.

Note on 'The Tales of the Heike'

These four stories are selections from *The Tales of the Heike*, a famous Japanese epic narrative of the twelfth century, dealing with the struggle for power between the Taira family (Heike) and the Minamoto family (Genji). The tales were for long recited by ballad-singers to the accompaniment of the lute, but later were collected in written form in twelve volumes. Part history and part legend they tell of the power and glory of the Heike under its brilliant but ruthless leader Kiyomori, and its tragic defeat and annihilation by the Genji under Yoritomo and his brother Yoshitsune. The heroic, moving, and often cruel deeds of heroes and heroines, which marked the long struggle between the two great clans, are known to every Japanese child and form part of the repertoire of the Japanese traditional theatres of Noh and Kabuki.

The present selection opens with a reference to the 'cloistered Emperor', and this perhaps needs a word of explanation. After the introduction of Chinese civilization into Japan, it became the practice of reigning monarchs to abdicate in favour of their successors and retire into private life to devote themselves to religious meditation. As time went on, there were Emperors who wished to retain the power of the throne, but at the same time to be relieved of its responsibilities. They followed the practice of abdication, but in reality continued to direct the affairs of state from their seclusion. Such Emperors were known as cloistered Emperors.